I0015763

Swift Cookbook

Proven recipes for developing robust iOS applications
with Swift 5.9

Keith Moon

Chris Barker

Daniel Bolella

Nathan Lawlor

Swift Cookbook

Copyright © 2024 Packt Publishing

All rights reserved. No part of this book may be reproduced, stored in a retrieval system, or transmitted in any form or by any means, without the prior written permission of the publisher, except in the case of brief quotations embedded in critical articles or reviews.

Every effort has been made in the preparation of this book to ensure the accuracy of the information presented. However, the information contained in this book is sold without warranty, either express or implied. Neither the authors, nor Packt Publishing or its dealers and distributors, will be held liable for any damages caused or alleged to have been caused directly or indirectly by this book.

Packt Publishing has endeavored to provide trademark information about all of the companies and products mentioned in this book by the appropriate use of capitals. However, Packt Publishing cannot guarantee the accuracy of this information.

Group Product Manager: Rohit Rajkumar

Publishing Product Manager: Nitin Nainani

Book Project Manager: Aishwarya Mohan

Senior Editor: Mudita S

Technical Editor: Reenish Kulshrestha

Copy Editor: Safis Editing

Proofreader: Mudita S

Indexer: Manju Arasan

Production Designer: Alishon Mendonca

DevRel Marketing Coordinators: Anamika Singh and Nivedita Pandey

First published: April 2015

Second edition: February 2021

Third edition: May 2024

Production reference: 2030724

Published by Packt Publishing Ltd.

Grosvenor House

11 St Paul's Square

Birmingham

B3 1RB, UK.

ISBN 978-1-80323-958-3

www.packtpub.com

To my wife and children, for their love, support, and inspiration. I love you all more than words can ever capture.

– Danny Bolella

I would like to thank my mother, Sofia, and my father, Graham, for their continuous support throughout my career and personal projects. Without them, I would not be where I am today.

– Nathan Lawlor

Contributors

About the authors

Keith Moon is an award-winning iOS developer, author, and speaker based in London. He has worked with some of the biggest companies in the world to create engaging and personal mobile experiences. Keith has been developing in Swift since its release, working on projects that are both fully Swift and mixed Swift and Objective-C. Keith has been invited to speak about Swift development at conferences from Moscow to Minsk and London.

Chris Barker is a Principal Software Engineer at Jaguar Land Rover, where he leads the Mobile Application Engineering Team across the business. With over 22 years of experience in the IT industry, Chris began his career developing .NET applications for the online retailer dabs.com (now BT Shop).

In 2014, Chris transitioned into mobile app development. Before joining Jaguar Land Rover, he worked on mobile apps for clients such as Louis Vuitton, L'Oréal Paris, SimplyBe, JD Williams, and Jacamo.

Chris is the co-host of NS Manchester, a local iOS developer meet-up in Manchester, UK. He has been involved in authoring, co-authoring, and reviewing books for Packt Publishing since 2020.

Daniel Bolella is a lead iOS engineer at a major financial services firm. With over a decade of experience under his belt, he's worked on everything from full stack web to mobile apps in a variety of industries, including financial, energy, and medical devices. He also enjoys writing articles and was the technical reviewer for the first and second editions of *SwiftUI Cookbook*.

Danny thanks his amazing wife and children, who give him the love, motivation, and drive to always be better. He thanks his parents, who lovingly encourage him to pursue his passions. He gives thanks to God, for always providing love and guidance. Lastly, he thanks all who have invested and mentored him, culminating in all he has become and achieved thus far.

Nathan Lawlor is a highly skilled iOS developer with many years of experience, initially starting his career as an apprentice in web development. Nathan has worked as a professional software developer in the home and fashion retail industry, with N Brown Group plc., and is now working in the automotive industry, with Jaguar Land Rover. He has published his own independent apps to the Apple App Store and regularly posts articles on his blog. Nathan has a passion for exploring new technologies and finding ways to improve code quality and best practices.

About the reviewers

Juan C. Catalan is a software engineer with more than 18 years of professional experience. He started mobile app development back in the days of iOS 3. Juan has worked as a professional iOS developer in many industries, including medical devices, financial services, real estate, document management, fleet tracking, and industrial automation. He has contributed to more than 30 published apps in the App Store, some of them with millions of users. Juan gives back to the iOS development community with technical talks, mentoring developers, and reviewing and authoring technical books. He is the author of *SwiftUI Cookbook*, *Third Edition* (Packt Publishing, 2023). Juan lives in Austin, Texas, with his wife Donna, where they spend time with their kids.

George MacKay-Shore is the lead engineer at AND Digital's Club Spärck in Halifax, UK, with over a decade of experience in the embedded systems, games, and enterprise software sectors, in both the public and private sector, and mobile, desktop, and web application spaces.

He is also an aspiring archer, a part-time comedian, and an avid walker, and when he's not reviewing books, he can be found learning new languages in the hope that he may one day use them!

Table of Contents

Preface   **xv**

1

Swift Fundamentals   1

Technical requirements	**2**
Writing your first code in Swift	**2**
Getting ready	2
How to do it…	2
There's more…	6
See also	7
Using the basic types – strings, ints, floats, and booleans	**7**
Getting ready	8
How to do it…	8
How it works…	10
There's more…	14
See also	14
Reusing code in functions	**14**
Getting ready	15
How to do it…	15
There's more…	17
See also	19
Encapsulating functionality in object classes	**19**
Getting ready	19
How to do it…	20
How it works…	22

There's more…	27
See also	28
Bundling values into structs	**28**
Getting ready	29
How to do it…	29
How it works…	30
There's more…	31
See also	32
Enumerating values with enums	**32**
Getting ready	33
How to do it…	33
How it works…	34
There's more…	35
See also	37
Passing around functionality with closures	**38**
Getting ready	38
How to do it…	39
How it works…	40
There's more…	42
See also	44
Using protocols to define interfaces	**44**
Getting ready	44

How to do it...   45   There's more...   47
How it works...   45   See also   49

2

Mastering the Building Blocks   51

Technical requirements   51   See also   81

Bundling variables into tuples 52   Changing your name with
Getting ready   52   a type alias   81
How to do it...   52   Getting ready   81
How it works...   53   How to do it...   81
There's more...   54   There's more...   82
See also   55   See also   83

Ordering your data with arrays 56   Getting property changing
Getting ready   56   notifications using
How to do it...   56   property observers   83
How it works...   59   Getting ready   84
There's more...   62   How to do it...   84
See also   63   How it works...   85
                                     There's more...   85
Containing your data in sets 63   See also   86
Getting ready   63
How to do it...   63   Extending functionality with
How it works...   65   extensions   86
See also   70   Getting ready   86
                                     How to do it...   86
Storing key-value pairs with   How it works...   87
dictionaries   70   There's more...   88
Getting ready   71   See also   89
How to do it...   72
How it works...   73   Controlling access with
There's more...   74   access control   89
See also   75   Getting ready   90
                                     How to do it...   94
Subscripts for custom types   76   How it works...   97
Getting ready   76   There's more...   101
How to do it...   77   See also   102
How it works...   80
There's more...   80

3

Data Wrangling with Swift   103

Technical requirements	103	Handling errors with try, throw, do, and catch	124
Making decisions with if/else	104		
Getting ready	104	Getting ready	124
How to do it...	104	How to do it...	124
How it works...	105	How it works...	126
There's more...	106	There's more...	127
See also	112	See also	131
Handling all cases with switch	112	Checking upfront with guard	131
Getting ready	112	Getting ready	132
How to do it...	112	How to do it...	132
How it works...	115	How it works...	134
See also	118	There's more...	135
		See also	136
Looping with for loops	118	Doing it later with defer	136
Getting ready	118	Getting ready	137
How to do it...	118	How to do it...	137
How it works...	119	How it works...	139
See also	121	There's more...	140
		See also	142
Looping with while loops	121	Bailing out with fatalError and precondition	142
Getting ready	121		
How to do it...	122	Getting ready	142
How it works...	122	How to do it...	143
There's more...	123	How it works...	144
See also	123	See also	145

4

Generics, Operators, and Nested Types   147

Technical requirements	147	How to do it...	149
Using generics with types	148	How it works...	152
Getting ready	148	There's more...	152

See also   154

Using generics with functions   154

Getting ready   154
How to do it...   154
How it works...   155
There's more...   156
See also   157

Using generics with protocols   157

Getting ready   157
How to do it...   157
How it works...   161
There's more...   165
See also   166

Using advanced operators   166

Getting ready   167
How to do it...   167
See also   170

Defining option sets   170

Getting ready   170
How to do it...   171
How it works...   171
See also   172

Creating custom operators   172

Getting ready   172
How to do it...   173
How it works...   175
There's more...   178
See also   180

Nesting types and namespacing   180

Getting ready   180
How to do it...   180
How it works...   182
There's more...   183
See also   183

5

Beyond the Standard Library   185

Technical requirements   185
Comparing dates with Foundation   186
Getting ready   186
How to do it…   186
How it works…   187
See also   189

Fetching data with URLSession   190
Getting ready   190
How to do it…   190
How it works…   191
See also   194

Working with JSON   194
Getting ready   195
How to do it...   196
There's more...   204

Working with XML   208
Getting ready   208
How to do it...   211
How it works...   216
There's more...   223
See also   225

6

Understanding Concurrency in Swift   227

Technical requirements	**228**	**Implementing the operation class**	**239**
Getting ready	228	Getting ready	239
How to do it...	229	How to do it...	239
How it works...	232	How it works...	246
See also	233	See also	247
Leveraging DispatchGroups	**233**	**Async/Await in Swift**	**247**
Getting ready	233	Getting ready	248
How to do it...	233	How to do it...	248
How it works...	237	How it works...	249
See also	239	See also	250

7

Building iOS Apps with UIKit   251

Technical requirements	**251**	Getting ready	274
Building an iOS app using UIKit and storyboards	**252**	How to do it...	275
		How it works...	280
Getting ready	252	There's more...	281
How to do it...	254	See also	282
How it works...	270	**UI testing with XCUITest**	**282**
There's more...	274	Getting ready	282
See also	274	How to do it...	283
		There's more...	286
Unit and integration testing with XCTest	**274**	See also	286

8

Building iOS Apps with SwiftUI   287

Technical requirements	**287**	How to do it…	288
Declarative syntax	**288**	How it works...	289
Getting ready	288	There's more...	290

See also   290
**Function builders, property
wrappers, and opaque return types   290**
Getting ready   290
How to do it…   291
There's more…   295
See also   296

Building simple views in SwiftUI   296
Getting ready   296

How to do it...   296
How it works...   304
There's more...   306
See also   309

Combine and data flow in SwiftUI   309
Getting ready   309
How to do it...   310
How it works...   313
See also   314

9

Getting to Grips with Combine   315

Technical requirements   315
Using Reactive Streams   316
Getting ready   316
How to do it...   316
How it works...   318
See also   318

Understanding Observable Objects 318
How to do it...   319
How it works...   321

See also   321

**Understanding publishers and
subscribers   321**
How to do it...   322
How it works...   324
See also   326

Combine versus Delegate pattern 326
How to do it...   327
How it works...   328

10

Using CoreML and Vision in Swift   329

Technical requirements   330
Getting ready   330
How to do it...   330
How it works...   333
There's more...   334
See also   334

**Using CoreML models to detect
objects in images   334**
Getting ready   335
How to do it...   335
How it works...   336
There's more...   337
See also   338

Building a video capture app	**338**	**Using CoreML and the Vision**	
Getting ready	338	**framework to detect objects in**	
How to do it...	338	**real time**	**342**
How it works...	340	Getting ready	342
There's more...	341	How to do it...	343
See also	342	How it works...	349
		See also	350

11

Immersive Swift with ARKit and Augmented Reality   351

Technical requirements	**351**	How it works...	361
Surface detection with ARKit	**352**	There's more…	362
Getting ready	352	**Using Reality Composer Pro**	
How to do it...	352	**for visionOS**	**362**
How it works...	354	Getting ready	363
There's more…	355	How to do it...	363
See also	357	How it works...	367
Using 3D models with ARKit	**358**	There's more…	368
Getting ready	358	See also	369
How to do it…	358		

12

Visualizing Data with Swift Charts   371

Technical requirements	**371**	How to do it...	375
Building a chart with data	**372**	How it works...	378
Getting ready	372	**Exploring chart marks**	
How to do it…	372	**and modifiers**	**379**
How it works...	374	How to do it...	380
See also	374	How it works...	382
Displaying multiple datasets	**374**	There's more...	383

Index   389

Other Books You May Enjoy   396

Preface

The Swift programming language, developed by Apple, has quickly become one of the most popular choices for building apps and services, specifically on the iOS platform. Swift, with its modern and expressive syntax, has an open source and robust library that can be used to create high-quality and efficient code for any scenario.

Swift 5.9 allows developers to take advantage of performant and responsive app-building techniques, using safe and clean code.

This book will guide you through the various features and capabilities that Swift offers, building up your knowledge, one step at a time, so that you can confidently build brilliant apps and services.

You will be given useful, easy-to-follow recipes to accomplish real-world tasks using Swift. Each recipe will build on the knowledge from previous topics that have been covered in the book.

Explore the limitless possibilities of the Swift programming language, empowering you to bring your ideas to life.

Who this book is for

If you are an aspiring developer looking to delve into the world of app development with the Swift programming language, then this book is for you. Whilst no solid experience with Swift is required, a basic understanding of programming concepts will be beneficial.

What this book covers

Chapter 1, *Swift Fundamentals*, introduces you to the basic concepts of Swift, its syntax, and the functionality of basic components. Additionally, you will be introduced to Apple's Xcode IDE and Swift Playgrounds, which provide developers with powerful tools to create, execute, and debug their code efficiently, while also preparing you to follow the recipes throughout this book. You will learn how to write your first lines of code using Swift and understand the various basic elements that the Swift programming language has to offer.

Chapter 2, Mastering the Building Blocks, teaches you how to create more complex structures, building on top of the basic components covered in the first chapter, as well as utilizing some more advanced functionality available in the Swift standard library. You will learn how to use arrays, dictionaries, tuples, and some more abstract concepts, such as extensions and property observers.

Chapter 3, Data Wrangling with Swift, explains the importance of making decisions within programming and how to alter the control flow of your code. You will learn how to conditionally execute code with the `if/else` and `switch` statements. Additionally, you will have the chance to explore other approaches with `for` and `while` loops, and even how to handle Swift errors with the `try`, `throw`, `do`, and `catch` statements.

Chapter 4, Generics, Operators, and Nested Types, covers two advanced features of Swift, which are generics and operators. These features will aid you in building functionality that is both flexible and well-defined. Moreover, you will understand how nested types can be beneficial by allowing you to group types logically, control access to your constructs, and use namespacing.

Chapter 5, Beyond the Standard Library, assesses the frameworks that sit outside of the standard library, specifically the Foundation framework. Learning how to take advantage of these broader functionalities will help you make full use of the Swift programming language, taking your projects to the next level.

Chapter 6, Understanding Concurrency in Swift, outlines the concept of concurrency in programming and how it can be used to increase the performance and responsiveness of your code. You will learn the fundamental approach of concurrency in Swift, with Dispatch Queues and Dispatch Groups. Then, you will look at the modern concurrency approach in Swift, with the Async/Await framework.

Chapter 7, Building iOS Apps with UIKit, starts your journey of learning how to build your very own iOS applications, using the traditional UIKit framework. You will understand how to use storyboards to create a user interface. Then, you will explore the approach of testing behaviors and visual components within your app, using the XCTest and XCUITest frameworks.

Chapter 8, Building iOS Apps with SwiftUI, welcomes you to the modern and declarative user interface framework that is SwiftUI. You will expand your knowledge of the concepts used to build an app, covered in the previous chapter, by using a simplified process to create beautiful, interactive, and dynamic user interfaces. Additionally, you will learn how to take advantage of live previews within Xcode when building your SwiftUI application.

Chapter 9, Getting to Grips with Combine, discusses reactive programming and how to use the Combine framework to handle events and data changes in a functional and simplified manner. You will understand how to use a wide range of operators to manipulate streams of data, as well as handling errors and canceling operations.

Chapter 10, Using CoreML and Vision in Swift, dives into the concepts of machine learning and also looks at the CoreML and Vision frameworks, covering how to process machine learning models to use within your apps. You will learn how to build an app for intelligent image recognition and adapt powerful techniques for a live video streamed on your device.

Chapter 11, *Immersive Swift with ARKit and Augmented Reality*, explores how to blend the virtual and physical worlds with 3D objects through the lens of your device's camera. You will learn the fundamentals of the ARKit framework and how to leverage the Augmented Reality tools available to create beautiful virtual scenes. Moreover, you will have the opportunity to create your first scene for visionOS.

Chapter 12, *Visualizing Data with Swift Charts*, shows you how to create visually appealing and easily understandable charts to represent data, using the Charts and SwiftUI frameworks. You will learn about the different types of chart markings, how to handle multiple datasets, and how to use visual modifiers.

To get the most out of this book

To follow along with the examples in this book, you will need a computer running macOS Sonoma (version 14.0) or greater. You also need an Apple ID to download and install Xcode 15 from the Mac App Store.

The code in this book has been tested using Swift 5.9 and should work with any new versions of Swift.

Software/hardware covered in the book	OS requirements
macOS	14.0+ (Sonoma)
Xcode	15

If you are using the digital version of this book, we advise you to type the code yourself or access the code via the GitHub repository (link available in the next section). Doing so will help you avoid any potential errors related to the copying and pasting of code.

Download the example code files

You can download the example code files for this book from GitHub at `https://github.com/PacktPublishing/Swift-Cookbook-Third-Edition`. If there's an update to the code, it will be updated on the existing GitHub repository.

We also have other code bundles from our rich catalog of books and videos available at `https://github.com/PacktPublishing/`. Check them out!

Conventions used

There are a number of text conventions used throughout this book.

`Code in text`: Indicates code words in text, database table names, folder names, filenames, file extensions, pathnames, dummy URLs, user input, and Twitter handles. Here is an example: "We define a new constant value by using the `let` keyword."

A block of code is set as follows:

```
let phrase: String = "The quick brown fox jumps over the lazy dog"
```

When we wish to draw your attention to a particular part of a code block, the relevant lines or items are set in bold:

```
let steve = Person.init(givenName: "Steven",
```

Any command-line input or output is written as follows:

```
<#What to append#> >> <#Where to append it#>
```

Bold: Indicates a new term, an important word, or words that you see on screen. For example, words in menus or dialog boxes appear in the text like this. Here is an example: "From here, in the toolbar on your Mac, click **File** > **New** > **Playground**…"

> **Tips or important notes**
> Appear like this.

Sections

In this book, you will find several headings that appear frequently (*Getting ready, How to do it..., How it works..., There's more...,* and *See also*).

To give clear instructions on how to complete a recipe, use these sections as follows:

Getting ready

This section tells you what to expect in the recipe and describes how to set up any software or any preliminary settings required for the recipe.

How to do it...

This section contains the steps required to follow the recipe.

How it works...

This section usually consists of a detailed explanation of what happened in the previous section.

There's more...

This section consists of additional information about the recipe in order to make you more knowledgeable about the recipe.

See also

This section provides helpful links to other useful information for the recipe.

Get in touch

Feedback from our readers is always welcome.

General feedback: If you have questions about any aspect of this book, mention the book title in the subject of your message and email us at customercare@packtpub.com.

Errata: Although we have taken every care to ensure the accuracy of our content, mistakes do happen. If you have found a mistake in this book, we would be grateful if you would report this to us. Please visit www.packtpub.com/support/errata, selecting your book, clicking on the Errata Submission Form link, and entering the details.

Piracy: If you come across any illegal copies of our works in any form on the Internet, we would be grateful if you would provide us with the location address or website name. Please contact us at copyright@packt.com with a link to the material.

If you are interested in becoming an author: If there is a topic that you have expertise in and you are interested in either writing or contributing to a book, please visit authors.packtpub.com.

Reviews

Please leave a review. Once you have read and used this book, why not leave a review on the site that you purchased it from? Potential readers can then see and use your unbiased opinion to make purchase decisions, we at Packt can understand what you think about our products, and our authors can see your feedback on their book. Thank you!

For more information about Packt, please visit packtpub.com.

Share Your Thoughts

Once you've read *Swift Cookbook*, we'd love to hear your thoughts! Scan the QR code below to go straight to the Amazon review page for this book and share your feedback.

https://packt.link/r/1803239581

Your review is important to us and the tech community and will help us make sure we're delivering excellent quality content.

Download a free PDF copy of this book

Thanks for purchasing this book!

Do you like to read on the go but are unable to carry your print books everywhere?

Is your eBook purchase not compatible with the device of your choice?

Don't worry, now with every Packt book you get a DRM-free PDF version of that book at no cost.

Read anywhere, any place, on any device. Search, copy, and paste code from your favorite technical books directly into your application.

The perks don't stop there, you can get exclusive access to discounts, newsletters, and great free content in your inbox daily

Follow these simple steps to get the benefits:

1. Scan the QR code or visit the link below

https://packt.link/free-ebook/9781803239583

2. Submit your proof of purchase
3. That's it! We'll send your free PDF and other benefits to your email directly

1

Swift Fundamentals

Since **Apple** announced the **Swift** programming language back in 2014 at the **Worldwide Developer Conference** (**WWDC**), it has gone on to become one of the fastest-growing programming languages.

Swift is a modern, general-purpose programming language that focuses on type safety and expressive and concise syntax. Positioned as a modern replacement for *Objective-C*, it has taken over from Apple's older language as the future of development across all their platforms.

Since open-sourcing Swift, Apple has provided support for running your Swift code on a whole host of platforms including **Linux**. Despite these alternative ways to use and write Swift code, the simplest is still on a *Mac* using Apple's **Xcode**.

In this chapter, we will look at the fundamentals of the Swift language and examine the syntax and functionality of the basic Swift component.

In this chapter, we will cover the following recipes:

- Writing your first code in Swift
- Using the basic types – strings, ints, floats, and booleans
- Reusing code in functions
- Encapsulating functionality in object classes
- Bundling values into structs
- Enumerating values with enums
- Passing around functionality with closures
- Using protocols to define interfaces

Technical requirements

We will walk you through setting up Xcode 15.0 and use this development environment unless otherwise stated. Xcode 15.0 can be downloaded from the Apple App Store on a Mac running the latest OS.

We will be using Swift 5.9. This version will also be more compatible with future versions of Swift, which means that code written now with Swift 5.9 can run alongside code written with future versions of Swift.

We will be using **Playgrounds in Xcode** to implement the recipes contained in this book unless otherwise stated. The benefit of using Xcode Playgrounds is its simplicity of quickly writing and compiling Swift syntax.

All the code for this chapter can be found in the book's GitHub repository at `https://github.com/PacktPublishing/Swift-Cookbook-Third-Edition/tree/main/Chapter%201`.

Writing your first code in Swift

In this recipe, we'll get you started with the Xcode **integrated development environment** (**IDE**) and get you ready to write your first lines of Swift code… buckle up!

Getting ready

For this recipe, you will need Xcode 15 or newer.

How to do it...

Once you have successfully downloaded Xcode from the Apple App Store, we'll need to launch the application:

1. Launch Xcode from the dock or via the Apple App Store, as shown in the following screenshot:

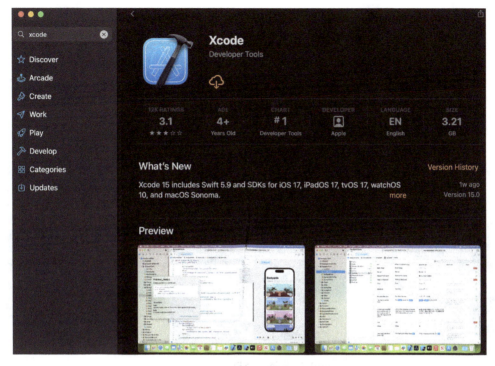

Figure 1.1 – Xcode in the App Store

2. You'll be presented with the following splash screen:

Figure 1.2 – Xcode splash screen

3. From here, in the toolbar on your Mac, click **File** > **New** > **Playground…**:

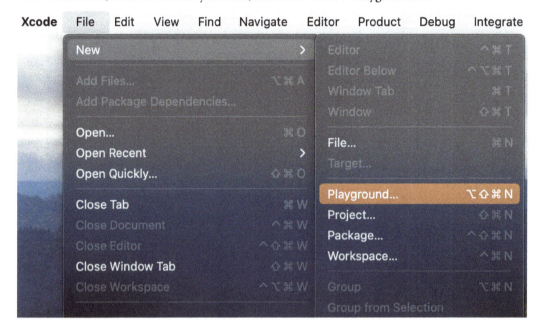

Figure 1.3 – Selecting Playground…

4. Select **Blank** from the iOS tab and press **Next**:

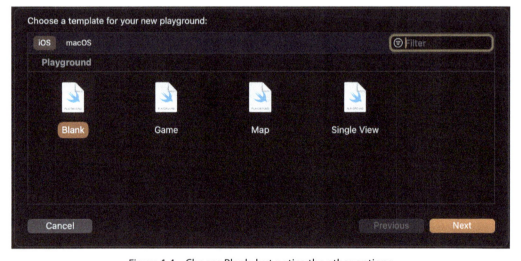

Figure 1.4 – Choose Blank, but notice the other options

5.   Choose a name and file location (this can be anything and anywhere you want) and press **Create**:

Figure 1.5 – Provide a name and location for our project

6.   You should now see the following playground:

Figure 1.6 – Our new playground

7.   Change the text to anything you want:

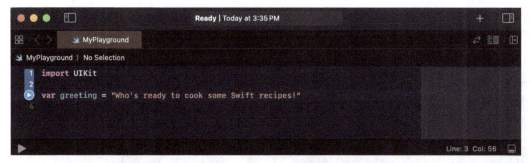

Figure 1.7 – Changing the string is as simple as replacing its text

8.   Now, press *play* at the bottom on the left just under the line numbers:

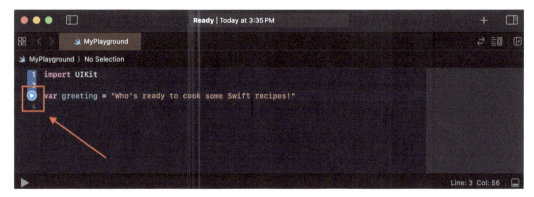

Figure 1.8 – Click the blue play button on the left

9. You should now see the output of your program in the right-hand column:

Figure 1.9 – Our output is now on the right

Congratulations, your first Swift program is now complete!

There's more...

If you put your cursor over the output column on the right-hand side, you will see two buttons, one that looks like an eye and another that is a rounded square:

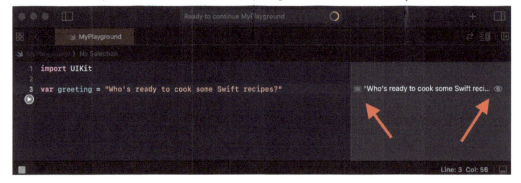

Figure 1.10 – Two icons are available on our output

Click on the eye button to get a *Quick Look* box of the output. This isn't particularly useful for a text string, but can be useful for more visual output, such as colors and views:

Figure 1.11 – The Quick Look box could quickly provide more details about our output

Click on the square button, and a box will be added inline, under your code, showing the output of the code. This can be useful if you want to see how the output changes as you change the code:

Figure 1.12 – Now we can see our output directly where it is called from

See also

For those curious about Swift's predecessor, the following link to Apple's documentation will be interesting: `https://developer.apple.com/library/archive/documentation/Cocoa/Conceptual/ProgrammingWithObjectiveC/Introduction/Introduction.html`.

Using the basic types – strings, ints, floats, and booleans

Many of the core operations in any programming language involve manipulating text and numbers, and determining `true` and `false` statements.

Let's learn how to accomplish these operations in Swift by looking at its basic types and learning how to assign constants and variables. In doing so, we will touch upon Swift's static typing and mutability system.

Getting ready

Just like we did in the previous recipe, open Xcode and create a new playground (again, call it what you want).

How to do it...

Let's start by writing some code that explores the basic types available to us in Swift.

1. Write the following code into your newly opened Swift Playground; we'll start with an example of Strings:

    ```
    let phrase: String = "The quick brown fox jumps over the lazy
    dog"
    ```

2. Now, let's add an example of an integer:

    ```
    let numberOfFoxes: Int = 1
    let numberOfAnimals: Int = 2
    ```

    The following is an example of how we use floating points in Swift:

    ```
    let averageCharactersPerWord: Float = (3+5+5+3+5+4+3+4+3) / 9
    print(averageCharactersPerWord) // 3.8888888

    /*
    phrase = "The quick brown ? jumps over the lazy ?"
    // Doesn't compile
    */
    ```

3. Now, add some code that handles concatenation in Swift and multiline expressions:

    ```
    var anotherPhrase = phrase
    anotherPhrase = "The quick brown 🐺 jumps over the lazy 🐶"
    print(phrase)
    // "The quick brown fox jumps over the lazy dog"
    print(anotherPhrase) // "The quick brown 🐺 jumps over the lazy
    🐶"

    var phraseInfo = "The phrase" + " has: "
    print(phraseInfo) // "The phrase has: "

    phraseInfo = phraseInfo + "\(numberOfFoxes) fox and \
    (numberOfAnimals) animals"
    print(phraseInfo)

    // "The phrase has: 1 fox and 2 animals"
    ```

```
print("Number of characters in phrase: \(phrase.count)")

let multilineExplanation = """
Why is the following phrase often used?
"The quick brown fox jumps over the lazy dog"
This phrase contains every letter in the alphabet.
"""

let phrasesAreEqual = phrase == anotherPhrase
print(phrasesAreEqual) // false

let phraseHas43Characters = phrase.count == 40 + 3
print(phraseHas43Characters) // true
```

4. Press the *play* button at the bottom of the window to run the playground and verify that Xcode doesn't show any errors.

Your playground should look like the following screenshot, with an output for each line in the timeline on the right-hand side and printed values in the console at the bottom:

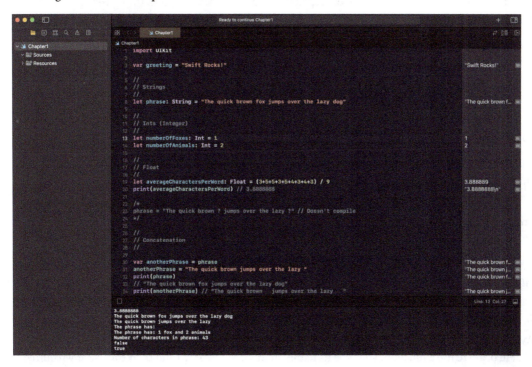

Figure 1.13 – Output in Swift Playground

How it works...

Let's step through the preceding code line by line to understand it.

In the following line of code, we are assigning some text to a constant value:

```
let phrase: String = "The quick brown fox jumps over the lazy dog"
```

We define a new constant value by using the `let` keyword, and we give that constant a name: `phrase`.

The colon (`:`) shows that we want to define what type of information we want to store in the constant, and that type is defined after the colon.

In this case, we want to assign a `String` type (`String` is how most programming languages refer to text).

The = sign indicates that we are assigning a value to the constant we have defined, and `"The quick brown fox jumps over the lazy dog"` is a `String` literal, which means that it's an easy way to construct a string.

Any text contained within `" "` marks is treated as a `String` literal by Swift.

We are assigning the `String` literal on the right-hand side of the = sign to the constant on the left-hand side of the = sign.

Next, we are assigning two more constants, but this time, they are of the `Int` type, or integers:

```
let numberOfFoxes: Int = 1
let numberOfAnimals: Int = 2
```

Rather than assigning a value directly, we can assign the outcome from a mathematical expression to the constant.

This constant is a `Float` type, or floating-point number:

```
let averageCharactersPerWord: Float = (3+5+5+3+5+4+3+4+3) / 9
```

In other words, it can store fractions rather than integers. Notice that in the timeline on the right of this line, the value is displayed as 3.88889.

The `print` function allows us to see the output from any expression printed to the console or displayed in the playground:

```
print(averageCharactersPerWord)
```

We will cover functions in a later recipe, *Reusing code in functions*, but for now, all you need to know is that in order to use a function, you type its name (in this case, `print`) and then enclose any required input to the function within brackets, `()`.

When our code calls this function, the timeline to the right of the code displays the output of the statement as 3.88888, which differs from the line above it.

The actual value of the mathematical expression we performed is 3.88888888... with an infinite number of 8s. However, the `print` function has rounded this up to just five decimal places and rounded it in a different way than the timeline for the line above.

This potential difference between the true value of a floating-point number and how it's represented by the Swift language is important to remember when dealing with floats.

Next, you'll see the following lines colored gray:

```
/*
phrase = "The quick brown ? jumps over the lazy ?" // Doesn't compile
*/
```

The playground doesn't produce an output for these lines because they are comments. The /* syntax before the line of code and the */ syntax after the line of code denotes that this is a comment block and, therefore, Swift should ignore anything typed in this block.

Remove /* and */ and you'll see that // Doesn't compile is still colored gray. This is because // also denotes a comment. Anything after this on the same line is also ignored.

If you now try and run this code, Xcode will tell you that there is a problem with this line, so let's look at the line to determine the issue.

On the left-hand side of the = sign, we have `phrase`, which we declared earlier, and now we are trying to assign a new value to it.

We can't do this because we defined `phrase` as a constant using the `let` keyword. We should only use `let` for things we know will not change.

This ability to define something as unchanging, or immutable, is an important concept in Swift, and we will revisit it in later chapters.

If we want to define something that can change, we declare it as a variable using the `var` keyword as follows:

```
var anotherPhrase = phrase
```

Since `anotherPhrase` is a variable, we can assign a new value to it:

```
anotherPhrase = "  The quick brown 🦊 jumps over the lazy 🐶"
```

Strings in Swift are fully Unicode compliant, so we can have some fun and use emojis instead of words.

Now, let's print out the values of our strings to see what values they hold:

```
print(phrase)
// "The quick brown fox jumps over the lazy dog"
print(anotherPhrase)
// "The quick brown 🦊 jumps over the lazy 🐶"
```

In this section, up till this point, we have done the following:

- Defined a string called phrase
- Defined a string called anotherPhrase as having the same value as phrase
- Changed the value of anotherPhrase
- Printed the value of phrase and anotherPhrase

In our output, we see that only anotherPhrase prints the new value that was assigned, even though the values of phrase and anotherPhrase were initially the same.

Although phrase and anotherPhrase had the same value, they do not have an intrinsic connection; so, when anotherPhrase is assigned a new value, this does not affect phrase.

Strings can be easily combined using the + operator:

```
var phraseInfo = "The phrase" + " has: "
print(phraseInfo) // "The phrase has: "
```

The preceding code gives the result you would expect; the strings are concatenated.

You will often want to create strings by including values derived from other expressions. We can do this with string interpolation:

```
phraseInfo = phraseInfo + "\(numberOfFoxes) fox and \(numberOfAnimals)
animals"
print(phraseInfo) // "The phrase has: 1 fox and 2 animals"
```

The values inserted after \(and before) can be anything that can be represented as a string, including other strings, integers, floats, or expressions.

We can also use expressions with string interpolation, such as displaying the number of characters in a string:

```
print("Number of characters in phrase: \(phrase.count)")
```

Strings in Swift are collections, which are containers of elements; in this case, a string is a collection of characters.

We will cover collections in more depth in *Chapter 2*, *Mastering the Building Blocks*, but for now, it's enough to know that your collections can tell you how many elements they contain through their count property.

We use this to output the number of characters in the phrase.

Multiline string literals can be defined using " " " at the beginning and end of the string:

```
let multilineExplanation = """
```

> **Interesting note**
>
> Why is the phrase, "The quick brown fox jumps over the lazy dog" often used in code?
>
> This phrase contains every letter in the alphabet!

The contents of the multiline string must be on a separate line from the start and end with signifiers. Within a multiline string literal, you can use a single quote character (") without needing to use an additional escape character, as you would with a single-line string literal.

Boolean or Bool values represent either true or false. In the next line, we evaluate the value of a Boolean expression and assign the result to the phrasesAreEqual constant:

```
let phrasesAreEqual: Bool = phrase == anotherPhrase
print(phrasesAreEqual) // false
```

In the preceding code, the equality operator, ==, compares the values on its left and right and evaluates to true if the two values are equal, and false otherwise.

As we discussed earlier, although we assigned anotherPhrase the value of phrase initially, we then assigned a new, different value to anotherPhrase; therefore, phrase and anotherPhrase are not equal and the expression assigns the value of false.

Each side of the == operator can be any expression that evaluates to match the type of the other side, as we do with the following code:

```
let phraseHas43Characters: Bool = phrase.characters.count == 40 + 3
print(phraseHas43Characters) // true
```

In this case, the character count of phrase equals 43. Since 40 + 3 also equals 43, the constant is assigned the value of true.

There's more...

During this recipe, we defined a number of constants and variables, and when we did this, we also explicitly defined their type. For example, consider the following line of Swift code:

```
let clearlyAString: String = "This is a string literal"
```

Swift is a statically typed language. This means any constant or variable that we define has to have a specific type, and once defined, it cannot be changed to a different type.

However, in the preceding line of code, the `clearlyAString` constant is clearly a string! The right-hand side of the expression is a string literal, and therefore, we know that the left-hand side will be a string.

More importantly, the Swift compiler also knows this (a compiler is the program that turns Swift code into machine code).

Swift is all about being concise, so since the type can be inferred by the compiler, we do not need to explicitly state it.

Instead of the preceding code, we can use the following code and it will still run, even though we didn't specify the type:

```
let clearlyAString = "This is a string literal"
```

In fact, all the type declarations that we have made so far can actually be removed.

Go back through the code we have already written and remove all type declarations (`:String`, `:Int`, `:Float`, and `:Bool`), as all these can be inferred.

Now, run the playground to confirm that this is still valid Swift code.

See also

Further information regarding these base types in Swift can be found in Apple's documentation of the Swift language:

Integers, Floats, and Booleans: `http://swiftbook.link/docs/the-basics`

Strings and characters: `http://swiftbook.link/docs/strings`

Reusing code in functions

Functions are a building block of almost all programming languages, allowing functionality to be defined and reused.

Swift's syntax provides an expressive way to define your functions, creating concise and readable code.

In this recipe, we will run through the different types of functions we can create and understand how to define and use them.

Getting ready

In this recipe, we can use the playground from the previous recipe. Don't worry if you didn't work through the previous recipe, as this one will contain all the code you need.

How to do it...

Let's look at how functions are defined in Swift:

```
func nameOfFunction(parameterLabel1 parameter1: ParameterType1,
parameterLabel2 parameter2: ParameterType2,...) -> OutputType {
    // Function's implementation
    // If the function has an output type,
    // the function must return a valid value return output
}
```

Let's look at this in more detail to see how a function is defined:

- func: This indicates that you are declaring a function.

- nameOfFunction: This will be the name of your function and, by convention, is written in camel case (this means that each word, apart from the first, is capitalized and all spaces are removed).

  The name should describe what the function does and should provide some context to the value returned by the function, if one is returned. This will also be how you will invoke the method from elsewhere in your code, so bear that in mind when naming it.

- parameterLabel1 parameter1: ParameterType1: This is the first input, or parameter, into the function.

  You can specify as many parameters as you like, separated by commas. Each parameter has a parameter name (parameter1) and type (ParameterType1). The parameter name is how the value of the parameter will be made available to your function's implementation.

  You can optionally provide a parameter label in front of the parameter name (parameterLabel1) that will be used to label the parameter when your function is used (at the call site).

- -> OutputType: This indicates that the function returns a value and indicates the type of that value. If no value is returned, this can be omitted.

- { }: The curly brackets indicate the start and end of the function's implementation; anything within them will be executed when the function is called.

- return output: If the function returns a value, you type return and then specify the value to return. This ends the execution of the function; any code written after the return statement is not executed.

Now, let's put our learning about functions into action.

Imagine that we are building a contacts app to hold the details of your family and friends, and we want to create a string of a contact's full name.

Let's explore some of the ways in which functions can be used:

```swift
// Input parameters and output
func fullName(givenName: String,
              middleName: String,
              familyName: String) -> String {
    return "\(givenName) \(middleName) \(familyName)"
}
```

The preceding function takes three string parameters and outputs a string that puts all these together with spaces in between.

The only thing this function does is take inputs and produce an output without causing any side effects; this type of function is often called a **pure function**.

To call this function, we enter the name of the function followed by the input parameters within brackets, `()`, where each parameter value is preceded by its label:

```swift
let myFullName = fullName(givenName: "Mandy",
                          middleName: "Mary",
                          familyName: "Barker")
print(myFullName) // Mandy Mary Barker
```

Since the function returns a value, we can assign the output of this function to a constant or a variable, just like any other expression.

The next function takes the same input parameters, but its purpose is not to return a value. Instead, it prints out the parameters as one string separated by spaces:

```swift
// Input parameters, with a side effect and no output
func printFullName(givenName: String,
                   middleName: String,
                   familyName: String) {
    print("\(givenName) \(middleName) \(familyName)")
}
```

We can call this function in the same way as the preceding function, although it can't be assigned to anything since it doesn't have a return value:

```swift
printFullName(givenName: "Mandy",
              middleName: "Mary",
              familyName: "Barker")
```

The following function takes no parameters as everything it needs to perform its task is contained within it, although it does output a string.

This function calls the `fullName` function we defined earlier, taking advantage of its ability to produce a full name when given the component names.

Reusing functionality is the most useful feature that functions provide. Let us see how to take advantage of this feature through the following code:

```
func fullName() -> String {
    return fullName(givenName: "Mandy",
                    middleName: "Mary",
                    familyName: "Barker")
}
```

Since `fullName` takes no parameters, we can execute it by entering the function name followed by empty brackets, `()`, and since it returns a value, we can assign the outcome of `fullName` to a variable:

```
let personsFullName = fullName()
```

Our final example takes no parameters and returns no value:

```
// No inputs, no output
func printFullName() {
    let personsFullName = fullName()
    print(personsFullName)
}
```

You can call this function in the same way as the previous functions with no parameters, and there is no return value to assign:

```
printFullName()
```

As you can see from the preceding example, having input parameters and providing an output value are not required when defining a function.

There's more...

Now, let's look at a couple of ways of making your use of functions more expressive and concise.

Default parameter values

One convenience in Swift is the ability to specify default values for parameters. These allow you to omit the parameter when calling, as the default value will be provided instead.

Let's use the same example as earlier in this recipe, where we are creating a contacts app to hold information about our family and friends.

Many of your family members are likely to have the same family name as you, so we can set the family name as the default value for that parameter. Therefore, the family name only needs to be provided if it is different from the default.

Enter the following code into a playground:

```
func fullName(givenName: String,
              middleName: String,
              familyName: String = "Pendlebury") -> String {

    return "\(givenName) \(middleName) \(familyName)"
}
```

Defining a default value looks similar to assigning a value to the familyName: String = "Pendlebury" parameter.

When calling the function, the parameter with the default value does not have to be given:

```
let chris = fullName(givenName: "Chris",
                     middleName: "Brian",
                     familyName: "Barker")
let madeleine = fullName(givenName: "Madeleine",
                     middleName: "Rose",
                     familyName: "Barker")
let mandy = fullName(givenName: "Mandy",
                     middleName: "Mary")

print(chris) // Chris Brian Barker
print(madeleine) // Madeleine Rose Barker
print(mandy) // Mandy Mary Pendlebury
```

Parameter overloading

Swift supports parameter overloading, which allows for functions to have the same name and only be differentiated by the parameters that they take.

Let's learn more about parameter overloading by entering the following code into a playground:

```
func combine(_ string1: String, _ string2: String) -> String {
    return "\(string1) \(string2)"
}
```

```
func combine(_ integer1: Int, _ integer2: Int) -> Int {
    return integer1 + integer2
}

let combinedString = combine("Madeleine", "Barker")
let combinedInt = combine(6, 10)

print(combinedString) // Madeleine Barker
print(combinedInt) // 16
```

Both the preceding functions have the name combined, but one takes two strings as parameters, and the other takes two integers.

Therefore, when we call the function, Swift knows which implementation we intended by the values we pass as parameters.

We've introduced something new in the preceding function declarations under *Default parameter values*: anonymous parameter labels such as `_ givenName: String`.

When we declare the parameters, we use an underscore, `_`, for the parameter label. This indicates that we don't want a parameter name shown when calling the function. This should only be used if the purpose of the parameters is clear without the labels.

See also

Further information about functions can be found at `https://docs.swift.org/swift-book/documentation/the-swift-programming-language/functions/`.

Encapsulating functionality in object classes

Object-oriented programming (OOP) is a programming paradigm common to most software development. At its core is the `object` class. Objects allow us to encapsulate data and functionality, which can then be stored and passed around.

In this recipe, we will build some class objects, to break down their components, and understand how they are defined and used.

Getting ready

In this recipe, we can use the playground from the previous recipe. Don't worry if you didn't work through the previous recipe, as this one will contain all the code you need.

How to do it...

Let's write some code to create and use class objects, and then we will walk through what the code is doing:

1. First, let's create a `Person` class object:

    ```
    class Person {
    }
    ```

2. Within the curly brackets, { }, add three constants representing the person's name, and one variable representing their country of residence:

    ```
    let givenName: String
    let middleName: String
    let familyName: String
    var countryOfResidence: String = "UK"
    ```

3. Below the properties, yet still within the curly brackets, add an initialization method for our `Person` object:

    ```
    init(givenName: String, middleName: String, familyName: String)
    {
        self.givenName = givenName
        self.middleName = middleName
        self.familyName = familyName
    }
    ```

4. Now, add a variable as a property of the class, with a computed value:

    ```
    var displayString: String {
            return "\(self.fullName()) - Location: \(self.
    countryOfResidence)"
        }
    ```

5. Add a function within the `Person` object that returns the person's full name:

    ```
    func fullName() -> String {
        return "\(givenName) \(middleName) \(familyName)"
    }
    ```

6. Next, create a `Friend` object that extends the functionality of the `Person` object:

    ```
    final class Friend: Person {
    }
    ```

7. Within the `Friend` class object, add a variable property to hold details of where the user met the friend, and override the display string property to customize its behavior for `Friend` objects:

```
var whereWeMet: String?
override var displayString: String {
        let meetingPlace = whereWeMet ?? "Don't know where
we met"
        return "\(super.displayString) - \(meetingPlace)"
}
```

8. In addition to the `Friend` object, create a `Family` object that extends the functionality of the `Person` object:

```
final class Family: Person {
}
```

9. Add a `relationship` property to the `Family` object and create an initializer method to populate it in addition to the other properties from `Person`:

```
final class Family: Person {
    let relationship: String
            init(givenName: String,
        middleName: String,
        familyName: String = "Barker",
        relationship: String) {

        self.relationship = relationship
        super.init(givenName: givenName,
                   middleName: middleName,
                   familyName: familyName)
    }
}
```

10. Give the `Family` object a custom `displayString` method that includes the value of the `relationship` property by adding this code within the `Family` object definition (within the curly brackets):

```
override var displayString: String {
        return "\(super.displayString) - \(relationship)"
}
```

11. Finally, create instances of the new objects and print the display string to see how its value differs:

```
let steve = Person(givenName: "Steven",
                   middleName: "Paul",
                   familyName: "Jobs")
```

```
let sam = Friend(givenName: "Sam",
                 middleName: "Wow",
                 familyName: "Rowley")
sam.whereWeMet = "Work together at Jaguar Land Rover"

let maddie = Family(givenName: "Madeleine",
                    middleName: "Barker",
                    relationship: "Daughter")

let mark = Family(givenName: "Mark",
                  middleName: "David",
                  familyName: "Pendlebury",
                  relationship: "Brother-In-Law")
mark.countryOfResidence = "UK"

print(steve.displayString)
// Steven Paul Jobs

print(sam.displayString)
// Sam Wow Rowley - Work together at Jaguar Land Rover

print(maddie.displayString)
// Madeleine Barker - Daughter

print(mark.displayString)
// Mark David Pendlebury - Brother-In-Law
```

How it works...

Classes are defined with the `class` keyword. Class names start with a capital letter by convention, and the implementation of the class is contained (or *scoped*) within curly brackets:

```
class Person {
    //...
}
```

An object can have property values, which are contained within the object.

These properties can have initial values, as `countryOfResidence` does in the following code:

```
let givenName: String
let middleName: String
let familyName: String
var countryOfResidence: String = "UK"
```

However, bear in mind that constants (defined with `let`) cannot be changed once the initial value has been set:

If your class were to just have the preceding property definitions, the compiler would raise a warning, as `givenName`, `middleName`, and `familyName` are defined as non-optional strings. However, we have not provided any way to populate those values.

The compiler needs to know how the object will be initialized so that we can be sure that all the non-optional properties will indeed have values:

```
class Person {
    let givenName: String
    let middleName: String
    let familyName: String
    var countryOfResidence: String = "UK"

    init(givenName: String,
         middleName: String,
         familyName: String) {
        self.givenName = givenName
        self.middleName = middleName
        self.familyName = familyName
    }
    //...
}
```

The `init` method is a special method (functions defined within objects are called methods) that's called when the object is initialized. In the `Person` object of the preceding code, `givenName`, `middleName`, and `familyName` must be passed in when the object is initialized, and we assign those provided values to the object's properties.

The `self.` prefix is used to differentiate between the property and the value passed in, as they have the same name.

We do not need to pass in a value for `countryOfResidence` as this has an initial value. This isn't ideal though, as when we create a `Person` object, it will always have the `countryOfResidence` variable set to `"UK"`, and we will then have to change that value, if different, after initialization.

Another way to do this would be to use a default parameter value, as seen in the previous recipe. Amend the `Person` object initialization to the following:

```
class Person {
    let givenName: String
    let middleName: String
    let familyName: String
```

```
    var countryOfResidence: String

    init(givenName: String,
        middleName: String,
        familyName: String,
        countryOfResidence: String = "UK") {

        self.givenName = givenName
        self.middleName = middleName
        self.familyName = familyName
        self.countryOfResidence = countryOfResidence
    }
    //...
}
```

Now, you can provide a country of residence in the initialization or omit it to use the default value.

Next, let's look at the `displayString` property of our `Person` class:

```
class Person {
    //...
    var displayString: String {
        return "\(self.fullName()) - Location: \(self.
countryOfResidence)"
    }
    //...
}
```

This property declaration is different from the others. Rather than having a value assigned to it, it is followed by an expression contained within curly braces.

This is a computed property; its value is not static but is determined by the given expression every time the property is accessed. Any valid expressions can be used to compute the property but must return a value that matches the declared type of the property.

The compiler will enforce this, and you can't omit the variable type for computed properties. In constructing the preceding `return` value, we use `self.fullName()` and `self.countryOfResidence`.

As we did in the preceding `init` method, we use `self.` to show that we are accessing the method and property of the current instance of the `Person` object.

However, since `displayString` is already a property on the current instance, the Swift compiler is aware of this context and so those self-references can be removed:

```
var displayString: String {
    return "\(fullName()) - Location:\(countryOfResidence)"
}
```

Objects can do work based on the information they contain, and this work can be defined in methods.

Methods are just functions that are contained within classes and have access to all the object's properties. The `Person` object's `fullName` method is an example of this:

```
class Person {
    //...
    func fullName() -> String {
        return "\(givenName) \(middleName) \(familyName))"
    }
    //...
}
```

All the abilities of a function are available, which we explored in the last recipe, *Reusing code in functions*, including optional inputs and outputs, default parameter values, and parameter overloading.

Having defined a `Person` object, we want to extend the concept of `Person` to define a friend. A friend is also a person, so it stands to reason that anything a `Person` object can do, a `Friend` object can also do.

We model this inherited behavior by defining `Friend` as a subclass of `Person`. We define the class that our `Friend` class inherits from (or the *superclass*), after the class name, separated by `:`, as follows:

```
final class Friend: Person {
    var whereWeMet: String?
    //...
}
```

By inheriting from `Person`, our `Friend` object inherits all the properties and methods from its superclass. We can then add any extra functionality we require. In this case, we add a property for details of where we met this friend.

The `final` prefix tells the compiler that we don't intend for this class to be subclassed; it is the final class in the inheritance hierarchy. This allows the compiler to make some optimizations as it knows it won't be extended.

In addition to implementing new functionalities, we can override functionalities from the superclass using the `override` keyword:

```
final class Friend: Person {
    //...
    override var displayString: String {
        let meetingPlace = whereWeMet ?? "Don't know where we met"
        return "\(super.displayString) - \(meetingPlace)"
    }
}
```

In the preceding code, we override the `displayString` computed property from `Person` as we want to add the `"where we met"` information.

Within the computed property, we can access the superclass's implementation by calling `super.`, and then referencing the property or method.

Next, let's look at how we can customize how our subclasses are initialized:

```swift
final class Family: Person {
    let relationship: String
    init(givenName: String,
         middleName: String,
         familyName: String = "Barker",
         relationship: String) {

        self.relationship = relationship
        super.init(givenName: givenName,
                   middleName: middleName,
                   familyName: familyName)
    }
    //...
}
```

Our `Family` class also inherits from `Person`, but we want to add a `relationship` property, which should form part of the initialization. So, we can declare a new `init` that also takes a `relationship` string value.

That passed-in value is then assigned to the `relationship` property because the superclass's initializer is called.

With all our class objects defined, we can create instances of these objects and call methods and access properties of these objects:

```swift
let steve = Person(givenName: "Steven",
                   middleName: "Paul",
                   familyName: "Jobs")

let sam = Friend(givenName: "Sam",
                 middleName: "Wow",
                 familyName: "Rowley")
sam.whereWeMet = "Work together at Jaguar Land Rover"

let maddie = Family(givenName: "Madeleine",
                    middleName: "Barker",
                    relationship: "Daughter")
```

```
let mark = Family(givenName: "Mark",
                  middleName: "David",
                  familyName: "Pendlebury",
                  relationship: "Brother-In-Law")
mark.countryOfResidence = "US"

print(steve.displayString)
// Steven Paul Jobs

print(sam.displayString)
// Sam Wow Rowley - Work together at Jaguar Land Rover

print(maddie.displayString)
// Madeleine Barker - Daughter

print(mark.displayString)
// Mark David Pendlebury - Brother-In-Law
```

To create an instance of an object, we use the name of the object like a function, passing in any required parameters. This returns an object instance that we can then assign to a constant or variable.

When creating an instance, we are actually calling the object's `init` method, and you can do this explicitly, as follows:

```
let steve = Person.init(givenName: "Steven",
                        middleName: "Paul",
                        familyName: "Jobs")
```

However, to be concise, this is usually omitted.

There's more...

Class objects are **reference types**, which is a term that refers to the way they are stored and referenced internally. To see how these reference type semantics work, let's look at how an object behaves when it is modified:

```
class VideoGameReview {
    let videoGameTitle: String
    var starRating: Int // Rating out of 5

    init(videoGameTitle: String, starRating: Int) {
        self.videoGameTitle = videoGameTitle
        self.starRating = starRating
    }
```

```
}

// Write a review
let monkeyIslandReview = VideoGameReview(videoGameTitle: "The Secret
of Monkey Island", starRating: 4)

// Post it to social media
let referenceToReviewOnTwitter = monkeyIslandReview
let referenceToReviewOnFacebook = monkeyIslandReview

print(referenceToReviewOnTwitter.starRating) // 4
print(referenceToReviewOnFacebook.starRating) // 4

// Reconsider the review
monkeyIslandReview.starRating = 5

// The change is visible from anywhere with a reference to the object
print(referenceToReviewOnTwitter.starRating) // 5
print(referenceToReviewOnFacebook.starRating) // 5
```

In the preceding code, we have defined a VideoGameReview class object, created an instance of that VideoGameReview object, and then assigned that review to two separate constants.

As a class object is a reference type, it is a reference to the object that is stored in the constant, rather than a new copy of the object.

Therefore, when we reconsider our review, to give the classic game *The Secret of Monkey Island* five stars, we are changing the underlying object. All references that access that underlying object will receive the updated value when the starRating property is accessed.

See also

Further information about classes can be found at https://docs.swift.org/swift-book/documentation/the-swift-programming-language/classesandstructures.

Bundling values into structs

Class objects are great for encapsulating data and functionality within a unifying concept, such as a person, as they allow individual instances to be referenced. However, not everything is an object.

We may need to represent data that is logically grouped together, but there isn't much more than that. It's not more than the sum of its parts; it is the sum of its parts.

For this, there are structs. Short for structures, structs can be found in many programming languages. Structs are value types (as opposed to classes, which are reference types) and, as such, behave differently when passed around. In this recipe, we will learn how structs work in Swift, and when and how to use them.

Getting ready

In this recipe, we will build on top of the previous recipe, so open the playground you have used for the previous recipe. Don't worry if you didn't work through the previous recipe, as this one will contain all the code you need.

How to do it...

We have already defined a `Person` object as having three separate string properties relating to the person's name. However, these three separate strings don't exist in isolation from each other, as together they define a person's name.

Currently, if you want to retrieve a person's name, you have to access three separate properties and combine them. Let's tidy this up by defining a person's name as its own struct:

1. Create a `struct` called `PersonName`:

    ```swift
    struct PersonName {
    }
    ```

2. Add three properties to `PersonName`, for `givenName`, `middleName`, and `familyName`. Make the first two into constants, and the last one into a variable, as a family name can change:

    ```swift
    struct PersonName {
        let givenName: String
        let middleName: String
        var familyName: String
    }
    ```

3. Add a method to combine the three properties into a `fullName` string:

    ```swift
    func fullName() -> String {
        return "\(givenName) \(middleName) \(familyName)"
    }
    ```

4. Provide a method to change the `familyName` property and prefix this method with the `mutating` keyword:

    ```swift
    mutating func change(familyName: String) {
        self.familyName = familyName
    }
    ```

5. Create a PersonName struct, passing in the property values:

```
var duncansName = PersonName(givenName: "Duncan",
                             middleName: "Zowie",
                             familyName: "Jones")
```

How it works...

Defining a struct is very similar to defining an object class, and that is intentional. Much of the functionality available to a class is also available to a struct. Therefore, you will notice that aside from using the struct keyword instead of class, the definitions of a class and a struct are almost identical.

Within the PersonName struct, we have properties for the three components of the name and the fullName method we saw earlier to combine the three name components into a fullName string.

The method we created to change the familyName property has a new keyword that we haven't seen before, mutating:

```
mutating func change(familyName: String) {
    self.familyName = familyName
}
```

This keyword must be added to any method in a struct that changes a property of the struct.

This keyword is to inform anyone using the method that it will change or mutate the struct. Unlike class objects, when you mutate a struct, you create a copy of the struct with the changed properties. This behavior is known as **value-type semantics**.

To see this in action, let's first create a struct and then check that it behaves as we expect when we assign it to different values:

```
let duncansBirthName = PersonName(givenName: "Duncan",
                                  middleName: "Zowie",
                                  familyName: "Jones")
print(duncansBirthName.fullName()) // Duncan Zowie Jones

var duncansCurrentName = duncansBirthName
print(duncansCurrentName.fullName()) // Duncan Zowie Jones
```

So far, so good. We have created a PersonName struct, assigned it to a constant called duncansBirthName, and then assigned that constant to a variable called duncansCurrentName.

Now, let's see what happens when we mutate duncansCurrentName:

```
duncansCurrentName.change(familyName: "Bowie")
print(duncansBirthName.fullName()) // Duncan Zowie Jones
print(duncansCurrentName.fullName()) // Duncan Zowie Bowie
```

When we call the `mutating` method on the `duncansCurrentName` variable, only that variable is changed. This change is not reflected in `duncansBirthName`, even though these structs were once the same.

This behavior would be different if `PersonName` was an `object` class, and we explored that behavior in the previous recipe.

There's more...

We can use how this value-type behavior interacts with constants and variables to restrict unintended changes.

To see this in action, first, let's amend our `Person` class to our new `PersonName` struct:

```
class Person {
    let birthName: PersonName
    var currentName: PersonName
    var countryOfResidence: String

    init(name: PersonName,
        countryOfResidence: String = "UK") {
        birthName = name
        currentName = name
        self.countryOfResidence = countryOfResidence
    }

    var displayString: String {
        return "\(currentName.fullName()) - Location: \
(countryOfResidence)"
    }

}
```

We've added the `birthName` and `currentName` properties of our new `PersonName` struct type, and we initiate them with the same value when the `Person` object is created.

Since a person's birth name won't change, we define it as a constant, but their current name can change, so it's defined as a variable.

Now, let's create a new `Person` object:

```
var name = PersonName(givenName: "Duncan", middleName: "Zowie",
familyName: "Jones")
let duncan = Person(name: name)
print(duncan.currentName.fullName()) // Duncan Zowie Jones
```

Since our `PersonName` struct has value semantics, we can use this to enforce the behavior that we expect our model to have. We would expect to not be able to change a person's birth name, and if you try, you will find that the compiler won't let you.

As we discussed earlier, changing the family name mutates the struct, and so a new copy is made. However, we defined `birthName` as a constant, which can't be changed, so the only way we would be able to change the family name would be to change our definition of `birthName` from `let` to `var`:

```
duncan.birthName.change(familyName: "Moon") // Does not compile.
// Compiler tells you to change let to var
```

When we change `currentName` to have a new family name, which we can do since we defined it as `var`, it changes the `currentName` property but not the `birthName` property, even though these were assigned with the same value:

```
print(duncan.birthName.fullName()) // Duncan Zowie Jones
print(duncan.currentName.fullName()) // Duncan Zowie Jones
duncan.currentName.change(familyName: "Bowie")
print(duncan.birthName.fullName()) // Duncan Zowie Jones
print(duncan.currentName.fullName()) // Duncan Zowie Bowie
```

We have used a combination of objects and structs to create a model that enforces our expected behavior. This technique can help to reduce potential bugs in our code.

See also

Further information about structs can be found at `https://docs.swift.org/swift-book/documentation/the-swift-programming-language/classesandstructures`.

Enumerating values with enums

An **enumeration** is a programming construct that lets you define a value type with a finite set of options. Most programming languages have enumerations (usually abbreviated to **enums**), although the Swift language takes the concept further than most.

An example of an enum from the iOS/macOS SDK is `ComparisonResult`, which you would use when sorting items. When comparing for the purposes of sorting, there are only three possible results from a comparison:

- Ascending: The items are ordered in ascending order
- Descending: The items are ordered in descending order
- Same: The items are the same

There are a finite number of possible options for a comparison result; therefore, it's a perfect candidate for being represented by an enum:

```
enum ComparisonResult : Int {
    case orderedAscending
    case orderedSame
    case orderedDescending
}
```

Swift takes the enum concept and elevates it to a first-class type. As we will see, this makes enums a very powerful tool for modeling your information.

This recipe will examine how and when to use enums in Swift.

Getting ready

This recipe will build on top of the earlier recipes, so open the playground you have used for the previous recipes. Don't worry if you haven't tried out the previous recipes, as this one will contain all the code you need.

How to do it...

In the *Encapsulating functionality in object classes* recipe, we created a Person object to represent people in our model, and in the *Bundling values into structs* recipe, we made a PersonName struct to hold information about a person's name.

Now, let's turn our attention to a person's title (for example, Mr. or Mrs.), which precedes someone's full name. There are a small and finite number of common titles that a person may have; therefore, an enum is a great way to model this information:

1. Create an enum to represent a person's title:

    ```
    enum Title {
        case mr
        case mrs
        case mister
        case miss
        case dr
        case prof
        case other
    }
    ```

2. We define our enumeration with the enum keyword and provide a name for the enum. As with classes and structs, the convention is that this starts with a capital letter, and the implementation is defined within curly brackets. We define each enum option with the case keyword, and, by convention, these start with a lowercase character. Assign the mr case of our Title enum to a value:

```
let title1 = Title.mr
```

3. Enums can be assigned by specifying the enum type, then a dot, and then the case. However, if the compiler can infer the enum type, we can omit the type and just provide the case, preceded by a dot.

4. Define a constant value of the Title type and then assign a case to it with the type inferred:

```
let title2: Title
title2 = .mr
```

How it works...

In many programming languages, including *C* and *Objective-C*, enums are defined as a type definition on top of an integer, with each case being given a defined integer value. In Swift, enums do not need to represent integers under the hood.

In fact, they do not need to be backed by any type and can exist as their own abstract concepts. Consider the following example:

```
enum CompassPoint {
    case North, South, East, West
}
```

It doesn't make sense to map the compass points as integers, and in Swift, we don't have to.

For Title also, an Int-based enum doesn't seem appropriate; however, a String-based one may be. So, let's declare our enum to be String-based:

```
enum Title: String {
    case mr = "Mr"
    case mrs = "Mrs"
    case mister = "Master"
    case miss = "Miss"
    case dr = "Dr"
    case prof = "Prof"
    case other // Inferred as "other"
}
```

The enum's raw underlying type is declared after its name and a : separator. The raw types that can be used to back the enum are limited to types that can be represented as a literal.

This includes the following Swift base types:

- String

- Int

- Float

- Bool

Cases can be assigned a value of the raw type; however, certain types can be inferred and so do not need to be explicitly declared. For `Int`-backed enums, the inferred values are sequentially assigned starting at 0:

```
enum Rating: Int {
    case worst // Inferred as 0
    case bad    // Inferred as 1
    case average // Inferred as 2
    case good // Inferred as 3
    case best // Inferred as 4
}
```

For `String`-based enums, the inferred value is the name of the case, so the other case in our `Title` enum is inferred to be `other`.

We can get the underlying value of the `enum` in its raw type by accessing its `rawValue` property:

```
let title1 = Title.mr
print(title1.rawValue) // "Mr"
```

There's more...

As mentioned in the introduction to this recipe, Swift treats enums as a first-class type; therefore, they can have functionality that is not available to enums in most programming languages. This includes having computed variables and methods.

Methods and computed variables

Let's imagine that it is important for us to know whether a person's title relates to a professional qualification that the person holds.

Let's add a method to our `enum` to provide that information:

```
enum Title: String {
    case mr = "Mr"
    case mrs = "Mrs"
    case mister = "Master"
```

```
    case miss = "Miss"
    case dr = "Dr"
    case prof = "Prof"
    case other // Inferred as "other"

    func isProfessional() -> Bool {
        return self == Title.dr || self == Title.prof
    }
}
```

For the list of titles that we have defined, `Dr` and `Prof` relate to professional qualifications, so we have our method return `true` if `self` (the instance of the `enum` type this method is called on) is equal to the `dr` case, or equal to the `prof` case.

This functionality feels more appropriate as a computed property since whether `isProfessional` applies or not is intrinsic to the `enum` itself, and we don't need to do much work to determine the answer. So, let's change this into a property:

```
enum Title: String {

    case mr = "Mr"
    case mrs = "Mrs"
    case mister = "Master"
    case miss = "Miss"
    case dr = "Dr"
    case prof = "Prof"
    case other // Inferred as "other"

    var isProfessional: Bool {
        return self == Title.dr || self == Title.prof
    }

}
```

Now, we can determine whether a title is a professional title by accessing the computed property on it:

```
let loganTitle = Title.mr
let xavierTitle = Title.prof
print(loganTitle.isProfessional) // false
print(xavierTitle.isProfessional) // true
```

We can't store any additional information on an enum, over and above the enum value itself, but being able to define methods and computed properties that provide extra information about the enum is a really powerful option.

Associated values

Our `String`-based enum seems perfect for our title information, except that we have a case called `other`. If the person has a title that we hadn't considered when defining the enum, we can choose `other`, but that doesn't capture what the `other` title is.

In our model, we would need to define another property to hold the value given for `other`, but that splits our definition of `Title` over two separate properties, which could cause an unintended combination of values.

Swift enums have a solution for this situation: associated values. We can choose to associate a value with each enum case, allowing us to bind a non-optional string to our other case.

Let's rewrite our `Title` enum to use an associated value:

```
enum Title {
    case mr
    case mrs
    case mister
    case miss
    case dr
    case prof
    case other(String)
}
```

We have defined the `other` case to have an associated value by putting the value's type in brackets after the case declaration. We do not need to add associated values for every case. Each case declaration can have associated values of different types or none at all.

Now, let's look at how we assign an enum case with an associated type:

```
let mister: Title = .mr
let dame: Title = .other("Dame")
```

The associated value is declared in brackets after the case, and the compiler enforces that the type matches the type declared in our enum definition.

As we declared the `other` case to have a non-optional string, we are ensuring that a title of `other` cannot be chosen without providing details of what the `other` title is, and we don't need another property to fully represent `Title` in our model.

See also

Further information about enums can be found at `http://swiftbook.link/docs/enums`.

Passing around functionality with closures

Closures are also referred to as **anonymous functions**, and this is the best way to explain them. Closures are functions without a name and, like other functions, they can take a set of input parameters and can return an output.

Closures behave like other primary types. They can be assigned, stored, passed around, and used as input and output to functions and other closures.

In this recipe, we will explore how and when to use closures in our code.

Getting ready

We will continue to build on our contacts app example from earlier in this chapter, so you should use the same playground as in the previous recipes.

If, however, you are implementing this in a new playground, first add the relevant code from the previous recipes:

```swift
struct PersonName {
    let givenName: String
    let middleName: String
    var familyName: String

    func fullName() -> String {
        return "\(givenName) \(middleName) \(familyName)"
    }

    mutating func change(familyName: String) {
        self.familyName = familyName
    }
}

class Person {
    let birthName: PersonName
    var currentName: PersonName
    var countryOfResidence: String

    init(name: PersonName, countryOfResidence: String = "UK") {
        birthName = name
        currentName = name
        self.countryOfResidence = countryOfResidence
    }

    var displayString: String {
```

```
        return "\(currentName.fullName()) - Location: \
(countryOfResidence)"
    }
}
```

How to do it...

Now, let's define a number of types of closures, which we will then work through step by step:

1. Define a closure to print this author's details that takes `No input` and returns `no output`:

    ```
    // No input, no output
    let printAuthorsDetails: () -> Void = {
        let name = PersonName(givenName: "Chris",
                              middleName: "Brian",
                              familyName: "Barker")
        let author = Person(name: name)
        print(author.displayString)
    }
    printAuthorsDetails() // "Chris Brian Barker - Location: UK"
    ```

2. Define a closure that creates a `Person` object. The closure takes `No input`, but returns a `Person` object as the output:

    ```
    // No input, Person output
    let createAuthor: () -> Person = {
        let name = PersonName(givenName: "Chris",
                              middleName: "Brian",
                              familyName: "Barker")
        let author = Person(name: name)
        return author
    }
    let author = createAuthor()
    print(author.displayString) // "Chris Brian Barker - Location:
    UK"
    ```

3. Define a closure that prints a person's details, taking the three components of their name as `String inputs`, but returning `no output`:

    ```
    // String inputs, no output
    let printPersonsDetails: (String, String, String) -> Void = {
    (given,      middle,     family) in
            let name = PersonName(givenName: given,
                                  middleName: middle,
                                  familyName: family)
    ```

```
        let author = Person(name: name)
        print(author.displayString)
    }
    printPersonsDetails("Mandy", "Mary", "Barker")
    // "Mandy Mary Barker - Location: UK"
```

4. Finally, define a closure to create a person, taking the three name components as `String` inputs and returning a `Person` object as the output:

```
    // String inputs, Person output
    let createPerson: (String, String, String) -> Person = {
    (given,    middle,    family) in
        let name = PersonName(givenName: given,
                              middleName: middle,
                              familyName: family)
        let person = Person(name: name)
        return person
    }

    let felix = createPerson("Madeleine", "Rose", "Barker")
    print(felix.displayString) // "Madeleine Rose Barker - Location:
    UK"
```

How it works...

Let's take a look at the different types of closures we just implemented:

```
// No input, no output
let printAuthorsDetails: () -> Void = {
    let name = PersonName(givenName: "Chris",
                          middleName: "Brian",
                          familyName: "Barker")
    let author = Person(name: name)
    print(author.displayString)
}
printAuthorsDetails() // "Chris Brian Barker - Location: UK"
```

As a first-class type in Swift, closures can be assigned to constants or variables, and constants and variables need a type.

To define a closure's type, we need to specify the input parameter types and the output type, and for the closure in the preceding code, the type is `()  -> Void`. The `Void` type is another way of saying *nothing*, so this closure takes no inputs and returns nothing, and the closure's functionality is defined within the curly brackets, as with other functions.

Now that we have this closure defined and assigned to the `printAuthorsDetails` constant, we can execute it like other functions, but with the variable name instead of the function's name.

We can use the following closure, which will cause this author's details to be printed:

```
printAuthorsDetails() // "Chris Brian Barker - Location: UK"
```

The next closure type takes `No` input parameters, but returns a `Person` object, as you can see with the `()  -> Person` type definition:

```
// No input, Person output
let createAuthor: () -> Person = {
    let name = PersonName(givenName: "Chris",
                          middleName: "Brian",
                          familyName: "Barker")
    let author = Person(name: name)
    return author
}
let author = createAuthor()
print(author.displayString) // "Chris Brian Barker - Location: UK"
```

Since it has an output, the execution of the closure returns a value that can be assigned to a variable or constant. In the preceding code, we execute the `createAuthor` closure and assign the output to the `author` constant.

Since we defined the closure type as `()  -> Person`, the compiler knows that the `output` type is `Person`, and so the type of constant can be inferred.

Since we don't need to declare it explicitly, let's remove the type declaration:

```
let author = createAuthor()
print(author.displayString) // "Chris Brian Barker - Location: UK"
```

Next, let's take a look at a closure that takes input parameters:

```
// String inputs, no output
let printPersonsDetails: (String, String, String) -> Void = {
(given, middle, family) in
    let name = PersonName(givenName: given,
                          middleName: middle,
                          familyName: family)
    let author = Person(name: name)
    print(author.displayString)
}
```

You will remember, from the *Reusing code in functions* recipe, that we can define parameter labels, which determine how the parameters are referenced when the function is used, and parameter names, which define how the parameter is referenced from within the function.

In closures, these are defined a bit differently:

1. Parameter labels cannot be defined for closures, so, when calling a closure, the order and parameter type have to be used to determine what values should be provided as parameters:

    ```
    (String, String, String) -> Void
    ```

2. Parameter names are defined inside the curly brackets, followed by the `in` keyword:

    ```
    (given, middle, family) in
    ```

3. Putting it all together, we can define and execute a closure with inputs and an output, as follows:

    ```
    // String inputs, Person output
    let createPerson: (String, String, String) -> Person = {
    (given, middle, family) in
        let name = PersonName(givenName: given,
                              middleName: middle,
                              familyName: family)
        let person = Person(name: name)
        return person
    }
    ```

There's more...

We've seen how we can store closures, but we can also use them as method parameters. This pattern can be really useful when we want to be notified when a long-running task is completed.

Let's imagine that we want to save the details of our `Person` object to a remote database, maybe for backup or use on other devices.

We may want to be notified when this process has been completed, so we execute some additional code, perhaps printing a completion message, or updating some UI. While the actual saving implementation is outside the scope of this recipe, we can amend our `Person` class to allow this `save` functionality to be called, passing a closure to execute on completion.

Add a method to save to a remote database, taking in a completion `Handler`, and store it for subsequent execution:

```
class Person {
    //....
    var saveHandler: ((Bool) -> Void)?
```

```
func saveToRemoteDatabase(handler: @escaping (Bool) -> Void) {
    saveHandler = handler
    // Send person information to remove database
    // Once remote save is complete, it calls saveComplete(Bool)
    // We'll fake it for the moment, and assume the save is
    // complete.
    saveComplete(success: true)
}
func saveComplete(success: Bool) {
    saveHandler?(success)
}
}
```

We define an optional variable to hold on to `saveHandler` during the long-running save operation. Our closure will take a `Bool` value to indicate whether the save was a success:

```
var saveHandler: ((Bool) -> Void)?
```

Let's now define a method to save our `Person` object, which takes a closure as a parameter:

```
func saveToRemoteDatabase(handler: @escaping (Bool) -> Void) {
    saveHandler = handler
    // Send person information to remove database
    // Once remote save is complete, it calls saveComplete(Bool)
    // We'll fake it for the moment, and assume the save is complete.
    saveComplete(success: true)
}
```

Our function stores the given closure in the variable and then starts the process of saving to the remote database (the actual implementation of this is outside the scope of this recipe). This `save` process will call the `saveComplete` method when completed.

We added a modifier, `@escaping`, just before the closure type definition. This tells the compiler that, rather than using the closure within this method, we intend to store the closure and use it later. The closure will be escaping the scope of this method.

This modifier is needed to prevent the compiler from doing certain optimizations that would be possible if the closure was `nonescaping`. It also helps users of this method understand whether the closure they provide will be executed immediately or at a later time.

With the `save` operation complete, we can execute the `saveHandler` variable, passing in the `success` Boolean:

```
func saveComplete(success: Bool) {
    saveHandler?(success)
}
```

Since we stored the closure as an optional, we need to unwrap it by adding a ? character after the variable name. If `saveHandler` has a value, the closure will be executed; if it is nil, the expression is ignored.

Now that we have a function that takes a closure, let's see how we call it:

```
let fox = createPerson("Mandy", "Mary", "Barker")
fox.saveToRemoteDatabase(handler: { success in
    print("Saved finished. Successful: \(success)")
})
```

Swift provides a more concise way to provide closures to functions.

When a closure is the last (or only) parameter, Swift allows it to be provided as a trailing closure. This means the parameter name can be dropped and the closure can be specified after the parameter brackets. So, we can rewrite the preceding code with the following neater syntax:

```
let fox = createPerson("Mandy", "Mary", "Barker")
fox.saveToRemoteDatabase() { success in
    print("Saved finished. Successful: \(success)")
}
```

See also

Further information about closures can be found at `https://docs.swift.org/swift-book/documentation/the-swift-programming-language/closures/`.

Using protocols to define interfaces

Protocols are a way to describe the interface that a type provides. They can be thought of as a contract, defining how you can interact with instances of that type.

Protocols are a great way to abstract *what* something does from *how* it does it. As we will see in subsequent chapters, Swift adds functionalities to protocols that make them even more useful and powerful than in many other programming languages.

Getting ready

We will continue to build on examples from the previous recipes, but don't worry if you haven't followed these recipes yet as all the code you need is listed in the upcoming sections.

How to do it...

In the previous recipe, we added a method to our `Person` class that (given the full implementation) would save it to a remote database. This is a very useful functionality, and as we add more features to our app, there will likely be more types that we also want to save to a remote database:

1. Create a protocol to define how we will interface with anything that can be saved in this way:

    ```
    protocol Saveable {
        var saveNeeded: Bool { get set }
        func saveToRemoteDatabase(handler: @escaping (Bool) -> Void)
    }
    ```

2. Update our `Person` class so that it conforms to the `Saveable` protocol:

    ```
    class Person: Saveable {
        //....
        var saveHandler: ((Bool) -> Void)?
        var saveNeeded: Bool = true
        func saveToRemoteDatabase(handler: @escaping (Bool) -> Void)
    {
            saveHandler = handler
            // Send person information to remove database
            // Once remote save is complete, it calls
            // saveComplete(Bool)
            // We'll fake it for the moment, and assume the save is
            // complete.
            saveComplete(success: true)
        }
        func saveComplete(success: Bool) {
            saveHandler?(success)
        }
    }
    ```

How it works...

Protocols are defined with the `protocol` keyword, and the implementation is contained within curly brackets. As we have seen with other type definitions, it is conventional to begin a protocol name with a capital letter. It is also conventional to name a protocol as either something that the type is or something that it does. In this protocol, we are declaring that any type of implementation is `Saveable`.

Types conforming to this protocol have two parts of the interface to implement. Let's look at the first:

```
var saveNeeded: Bool { get set }
```

The `Saveable` protocol declares that anything implementing it needs to have a variable called `saveNeeded`, which is a `Bool` type.

This property will indicate that the information held in the remote database is out of date and a save is needed. In addition to the usual property declaration, a protocol requires us to define whether the property can be accessed (`get`) and changed (`set`), which is added in curly brackets after the type declaration.

Removing the set keywords makes it a read-only variable.

The second part of our `protocol` definition is to describe the method we can call to save the information to the remote database:

```
func saveToRemoteDatabase(handler: @escaping (Bool) -> Void)
```

This `func` declaration is exactly the same as other function declarations we have seen. However, the implementation of this function, which would have been contained in curly brackets, is omitted. Any type conforming to this protocol must provide this function and its implementation.

Now that we have defined our protocol, we need to implement the `Saveable` protocol on our `Person` class that we have been using throughout this chapter:

```
class Person: Saveable {
    //....
    var saveHandler: ((Bool) -> Void)?

    func saveToRemoteDatabase(handler: @escaping (Bool) -> Void) {
        saveHandler = handler
        // Send person information to remove database
        // Once remote save is complete, it calls
        // saveComplete(Bool)
        // We'll fake it for the moment, and assume the save is
        // complete.
        saveComplete(success: true)
    }

    func saveComplete(success: Bool) {
        saveHandler?(success)
    }
}
```

Conforming to a protocol looks similar to how a class inherits from another class, as we saw earlier in this chapter.

The protocol name is added after the type name, separated by :. By adding this conformance, the compiler will complain that our `Person` object doesn't implement part of the protocol, as we haven't declared a `saveNeeded` property. So, let's add that:

```
class Person: Saveable {
    //....
    var saveHandler: ((Bool) -> Void)?

    func saveToRemoteDatabase(handler: @escaping (Bool) -> Void) {
        saveHandler = handler
        // Send person information to remove database
        // Once remote save is complete, it calls
        // saveComplete(Bool)
        // We'll fake it for the moment, and assume the save is
        // complete.
        saveComplete(success: true)
    }

    func saveComplete(success: Bool) {
        saveHandler?(success)
    }
}
```

We'll add a default value of true since when an instance of this object is created, it won't be in the remote database, so it will need to be saved.

There's more...

Protocol conformance can be applied to classes, structs, enums, and even other protocols. The benefit of a protocol is that it allows an instance to be stored and passed without needing to know how it's implemented under the hood.

This provides many benefits, including testing using mock objects and changing implementations without changing how and where the implementations are used.

Let's add a feature to our app that lets us set a reminder for a contact's birthday, which we will also want to save to our remote database.

We can use protocol conformance to give our reminder the same consistent `save` functionality interface, even though a reminder may have a very different implementation for saving.

Let's create our `Reminder` object and have it conform to the `Saveable` protocol:

```
class Reminder: Saveable {
    var dateOfReminder: String // There is a better way to store
dates, but this will suffice currently.
```

```
        var reminderDetail: String // eg. Ali's birthday
        init(date: String, detail: String) {
            dateOfReminder = date
            reminderDetail = detail
        }
        var saveHandler: ((Bool) -> Void)?
        var saveNeeded: Bool = true


        func saveToRemoteDatabase(handler: @escaping (Bool) -> Void) {
    saveHandler = handler
            // Send reminder information to remove database
            // Once remote save is complete, it calls
            // saveComplete(success: Bool)
            // We'll fake it for the moment, and assume the save is
            // complete.
            saveComplete(success: true)
        }

        func saveComplete(success: Bool) {
            saveHandler?(success)
        }
    }
}
```

Our `Reminder` object conforms to `Saveable` and implements all the requirements.

We now have two objects that represent very different things and have different functionalities, but they both implement `Saveable`; therefore, we can treat them in a common way.

To see this in action, let's create an object that will manage the saving of information in our app:

```
class SaveManager {
    func save(_ thingToSave: Saveable) {
        thingToSave.saveToRemoteDatabase { success in
            print("Saved! Success: \(success)")
        }
    }
}


let maddie = createPerson("Madeleine", "Rose", "Barker")
// This closure was
// covered in the previous recipe

let birthdayReminder = Reminder(date: "08/06/2006", detail: "Maddie's
Birthday")
let saveManager = SaveManager()
```

```
saveManager.save(maddie)
saveManager.save(birthdayReminder)
```

In the preceding example, `SaveManager` doesn't know the underlying type that it is being passed, but it doesn't need to. It receives instances that conform to the `Saveable` protocol and, therefore, can use that interface to save each instance.

See also

Further information about protocols can be found at `https://docs.swift.org/swift-book/documentation/the-swift-programming-language/protocols/`.

2

Mastering the Building Blocks

The previous chapter explained the basic types that form the building blocks of the Swift language. In this chapter, we will build on this knowledge to create more complex structures, such as arrays and dictionaries, before moving on and looking at some of the little gems Swift offers, such as tuples and type aliases. Finally, we'll round off this chapter by looking at extensions and access control – both of which are key components that contribute to a sound yet efficient code base.

By the end of this chapter, you will be better equipped to organize and handle any data you'll want to work with in your apps, plus the ability to make the code that interacts with your data more flexible to your needs!

In this chapter, we will cover the following recipes:

- Bundling variables into tuples
- Ordering your data with arrays
- Containing your data in sets
- Storing key-value pairs with dictionaries
- Subscripts for custom types
- Changing your name with a type alias
- Getting property-changing notifications using property observers
- Extending functionality with extensions
- Controlling access with access control

Let's get started!

Technical requirements

All the code for this chapter can be found in this book's GitHub repository at https://github.com/PacktPublishing/Swift-Cookbook-Third-Edition/tree/main/Chapter%202.

Bundling variables into tuples

A **tuple** is a combination of two or more values that can be treated as one. If you have ever wished you could return more than one value from a function or method, without defining a new struct or class, you should find tuples very interesting.

Getting ready

Create a new playground and add the following statement:

```
import Foundation
```

This example uses one function from Foundation. We will delve into Foundation in more detail in *Chapter 5*, *Beyond the Standard Library*, but for now, we just need to import it.

How to do it...

Let's imagine that we are building an app that pulls movie ratings from multiple sources and presents them together, helping a user decide which movie to watch. These sources may use different rating systems, such as the following:

- The number of stars out of 5
- Points out of 10
- The percentage score

We want to normalize these ratings so that they can be compared directly and displayed side by side. We want all the ratings to be represented as the number of stars out of 5, so we will write a function that will return the number of whole stars out of 5. We will then use this to display the correct number of stars in our **user interface (UI)**.

Our UI also includes a label that will read **x Star Movie**, where **x** is the number of stars. It would be useful if our function returned both the number of stars and a string that we can put in the UI. We can use a tuple to do this. Let's get started:

1. Create a function to normalize the star ratings. The following function takes a rating and a total possible rating, and then returns a tuple of the normalized rating and a string to display in the UI:

    ```
    func normalizedStarRating(forRating rating: Float,
    ofPossibleTotal total: Float) -> (Int, String) {

    }
    ```

2. Inside the function, calculate the fraction of the total score. Then, multiply that by our normalized total score, 5, and round it up to the nearest whole number:

```
let fraction = rating / total
let ratingOutOf5 = fraction * 5
let roundedRating = round(ratingOutOf5) // Rounds to the nearest
// integer.
```

3. Still within the function, take the rounded fraction and convert it from `Float` into `Int`. Then, create the display string and return both `Int` and `String` as a tuple:

```
let numberOfStars = Int(roundedRating) // Turns a Float into an
Int
let ratingString = "\(numberOfStars) Star Movie"
return (numberOfStars, ratingString)
```

4. Call our new function and store the result in a constant:

```
let ratingAndDisplayString = normalisedStarRating(forRating: 5,
ofPossibleTotal: 10)
```

5. Retrieve the number of stars rating from the tuple and print the result:

```
let ratingNumber = ratingAndDisplayString.0
print(ratingNumber) // 3 - Use to show the right number of stars
```

6. Retrieve the display string from the tuple and print the result:

```
let ratingString = ratingAndDisplayString.1
print(ratingString) // "3 Star Movie" - Use to put in the label
```

With that, we have created and used a tuple.

How it works...

A tuple is declared as a comma-separated list of the types it contains, within brackets. In the preceding section, in *step 1*, you can see a tuple being declared as `(Int, String)`. The function, `normalizedStarRating`, normalizes the rating and creates `numberOfStars` as the closest round number of stars and `ratingString` as a display string. These values are then combined into a tuple by putting them, separated by a comma, within brackets – that is, `(numberOfStars, ratingString)` in *step 3*. This tuple value is then returned by the function.

Now, let's look at what we can do with that returned tuple value:

```
let ratingAndDisplayString = normalizedStarRating(forRating: 5,
ofPossibleTotal: 10)
```

Calling our function returns a tuple that we store in a constant called `ratingAndDisplayString`. We can access the tuple's components by accessing the numbered member of the tuple:

```
let ratingNumber = ratingAndDisplayString.0
print(ratingNumber) // 3 - Use to show the right number of stars

let ratingString = ratingAndDisplayString.1
print(ratingString) // "3 Star Movie" - Use to put in the label
```

> **Note**
>
> As is the case with most numbered systems in programming languages, the member numbering system starts with 0. The number that's used to identify a certain place within a numbered collection is called an **index**.

There is another way to retrieve the components of a tuple that can be easier to remember than the numbered index. By specifying a tuple of variable names, each value of the tuple will be assigned to the respective variable names. Due to this, we can simplify accessing the tuple values and printing the result:

```
let (nextNumber, nextString) = normalizedStarRating(forRating: 8,
ofPossibleTotal: 10)
print(nextNumber) // 4
print(nextString) // "4 Star Movie"
```

Since the numerical value is the first value in the returned tuple, this gets assigned to the `nextNumber` constant, while the second value, the string, gets assigned to `nextString`. These can then be used like any other constant, eliminating the need to remember which index refers to which value.

There's more...

As we mentioned previously, accessing a tuple's components via a number is not ideal, as we have to remember their order in the tuple to ensure that we are accessing the correct one. To provide some context, we can add labels to the tuple components, which can be used to identify them when they are accessed. Tuple labels are defined in a similar way to parameter labels, preceding the type and separated by `:`. Let's add labels to the function we created in this recipe and then use them to access the tuple values:

```
func normalizedStarRating(forRating rating: Float,
ofPossibleTotal total: Float) -> (starRating: Int, displayString:
String) {

  let fraction = rating / total
  let ratingOutOf5 = fraction * 5
  let roundedRating = round(ratingOutOf5) // Rounds to the nearest
integer
```

```
let numberOfStars = Int(roundedRating) // Turns a Float into an Int
let ratingString = "\(numberOfStars) Star Movie"
return (starRating: numberOfStars,
   displayString: ratingString)
}

let ratingAndDisplayString = normalizedStarRating(forRating: 5,
ofPossibleTotal: 10)
let ratingInt = ratingAndDisplayString.starRating
print(ratingInt) // 3 - Use to show the right number of stars
let ratingString = ratingAndDisplayString.displayString
print(ratingString) // "3 Stars" - Use to put in the label
```

As part of the function declaration, we can see the tuple being declared:

```
(starRating: Int, displayString: String)
```

When a tuple of that type is created, the provided values are preceded by the label:

```
return (starRating: numberOfStars, displayString: ratingString)
```

To access the components of the tuple, we can use these labels (although the number of indexes still works):

```
let ratingValue = ratingAndDisplayString.starRating print(ratingValue)
// 3 - Use to show the right number of stars

let ratingString = ratingAndDisplayString.displayString
print(ratingString) // "3 Stars" - Use to put in the label
```

Tuples are a convenient and lightweight way to bundle values together.

> **Tip**
> In this example, we created a tuple with two components. However, a tuple can contain any number of components.

See also

Further information about tuples can be found in Apple's documentation on the Swift language at https://docs.swift.org/swift-book/documentation/the-swift-programming-language/types.

Ordering your data with arrays

So far in this book, we have learned about many different Swift constructs – *classes*, *structs*, *enums*, *closures*, *protocols*, and *tuples*. However, it is rare to deal with just one instance of these on their own. Often, we will have many of these constructs, and we need a way to collect multiple instances and place them in useful data structures. Over the following few recipes, we will examine three collection data structures provided by Swift – that is, **arrays**, **sets**, and **dictionaries** (dictionaries are often called **hash tables** in other programming languages):

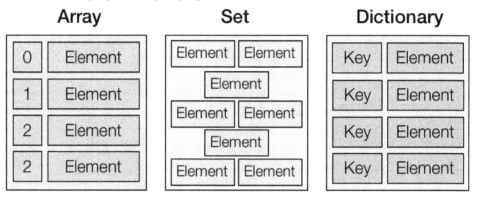

Figure 2.1 – A collection of data structures

While doing this, we will look at how to use them to store and access information, and then examine their relative characteristics.

Getting ready

First, let's investigate arrays, which are an ordered list of elements. We won't be using any components from the previous recipes, so you can create a new playground for this recipe.

How to do it...

Let's use an array to organize a list of movies to watch:

1. Create an array called `gamesToPlay`. This will hold our strings:

    ```
    var gamesToPlay = [String]()
    ```

2. Append three movies to the end of our movie list array:

    ```
    gamesToPlay.append("The Secret of Monkey Island")
    gamesToPlay.append("Half Life 2")
    gamesToPlay.append("Alien Isolation")
    ```

3. Print the names of each movie in the list, in turn:

```
print(gamesToPlay[0]) // "The Secret of Monkey Island"
print(gamesToPlay[1]) // "Half Life 2"
print(gamesToPlay[2]) // "Alien Isolation"
```

4. Print a count of the number of movies in the list so far:

```
print(gamesToPlay.count) // 3
```

5. Insert a new movie into the list so that it's the third one in it. Since arrays are zero-based, this is done at index 2:

```
gamesToPlay.insert("Breath of the Wild", at: 2)
```

6. Print the list count to check that it has increased by one, and print the newly updated list:

```
print (gamesToPlay.count) // 4
print(gamesToPlay)
// "The Secret of Monkey Island"
// "Half Life 2"
// "Breath of the Wild"
// "Alien Isolation"
```

7. Use the `first` and `last` array properties to access their respective values and print them:

```
let firstGameToPlay = gamesToPlay.first ?? ""
print(firstGameToPlay) // "The Secret of Monkey Island"
let lastGameToPlay = gamesToPlay.last ?? ""
print(lastGameToPlay as Any) // "Alien Isolation"
```

8. Use an index subscript to access the second movie in the list and print it. Then, set a new value to that same subscript. Once you've done that, print the list count to check the number of movies that haven't changed, and print the list to check that the second array element has changed:

```
let secondMovieToWatch = gamesToPlay[1]
print(secondMovieToWatch) // "Ghostbusters"

gamesToPlay[1] = "Half Life 2 (2004)"
print(gamesToPlay.count) // 4
print(gamesToPlay)
// "The Secret of Monkey Island"
// "Half Life 2 (2004)"
// "Breath of the Wild"
// "Alien Isolation"
```

9. Create a new array of spy movies by initializing it with some movies, using the array literal syntax:

```
let graphicAdventureGames: [String] = ["Monkey Island 2",
    "Loom",
    "Sam & Max"]
```

10. Combine the two arrays we have created, using the addition operator (+), and assign them back to the gamesToPlay variable. Then, print the array count so that it reflects the two lists combined, and print the new list:

```
gamesToPlay = gamesToPlay + graphicAdventureGames
print(gamesToPlay.count) // 7
print(gamesToPlay)
// "The Secret of Monkey Island"
// "Half Life 2 (2004)"
// "Breath of the Wild"
// "Alien Isolation"
// "Monkey Island 2"
// "Loom"
// "Sam & Max"
```

11. Now, use an array convenience initializer to create an array that contains three entries that are the same. Then, update each array element so that the rest of their movie titles are shown:

```
var batmanGames = Array<String>(repeating: "Batman: ", count: 3)
batmanGames[0] = batmanGames[0] + "Arkham Asylum"
batmanGames[1] = batmanGames[1] + "Arkham City"
batmanGames[2] = batmanGames[2] + "Arkham Knight"
print(batmanGames)
// Batman: Arkham Asylum
// Batman: Arkham City
// Batman: Arkham Knight
```

12. Let's replace part of our existing movie list with our batmanGames list, and then print the count and list:

```
gamesToPlay.replaceSubrange(2...4, with: batmanGames)
print(gamesToPlay.count) // 7
print(gamesToPlay)
// "The Secret of Monkey Island"
// "Half Life 2 (2004)"
// Batman: Arkham Asylum
// Batman: Arkham City
```

```
// Batman: Arkham Knight
// "Breath of the Wild"
// "Alien Isolation"
```

13. Lastly, remove the last movie in the list and check that the array count has reduced by one:

```
gamesToPlay.remove(at: 6)
print(gamesToPlay.count) // 6
print(gamesToPlay)
// "The Secret of Monkey Island"
// "Half Life 2 (2004)"
// Batman: Arkham Asylum
// Batman: Arkham City
// Batman: Arkham Knight
// "Breath of the Wild"
```

With that, we've looked at many ways we can create and manipulate arrays.

How it works...

When creating an array, we need to specify the type of elements that will be stored in the array. The array element type is declared in angular brackets as part of the array's type declaration. In our case, we are storing strings:

```
var gamesToPlay = Array<String>()
gamesToPlay.append("The Secret of Monkey Island")
gamesToPlay.append("Half Life 2")
gamesToPlay.append("Alien Isolation")
```

The preceding code uses a Swift language feature called **generics**, which can be found in many programming languages and will be covered in detail in *Chapter 4, Generics, Operators, and Nested Types*.

The append method of `Array` will add a new element to the end of the array. Now that we have put some elements in the array, we can retrieve and print those elements:

```
print(gamesToPlay[0]) // "The Secret of Monkey Island"
print(gamesToPlay[1]) // "Half Life 2"
print(gamesToPlay[2]) // "Alien Isolation"
```

Elements in an array are numbered with a zero-based index, so the first element in the array is at index 0, the second is at index 1, the third is at index 2, and so on. We can access the elements in the array using a subscript, in which we provide the index of the element we want to access. A subscript is specified in square brackets, after the array instance's name.

When an element is accessed using the index subscript, no check is done to ensure you have provided a valid index. In fact, if an index is provided that the array doesn't contain, this will cause a crash. Instead, we can use some index helper methods on `Array` to ensure that we have an index that is valid for this array. Let's use one of these helper methods to check an index that we know is valid for our array, and then another that we know is not valid:

```
let index5 = gamesToPlay.index(gamesToPlay.startIndex,
offsetBy: 5,
limitedBy: gamesToPlay.endIndex) print(index5 as Any) // Optional(5)

let index10 = gamesToPlay.index(gamesToPlay.startIndex,
offsetBy: 10,
limitedBy: gamesToPlay.endIndex)
print(index10 as Any) // nil
```

The `index` method lets us specify the index we want as an offset of the first index parameter, but as something that's limited by the last index parameter. This will return the valid index if it is within the bounds, or `nil` if it is not. By the end of the playground, the `gamesToPlay` array contains six elements, in which case retrieving index 5 is successful, but index `10` returns `nil`.

In *Chapter 3, Data Wrangling with Swift*, we will cover how to make decisions based on whether this index exists, but for now, it's just useful to know that this method is available.

Arrays have a `count` property that tells us how many elements they store. So, when we add an element, this value will change:

```
print(gamesToPlay.count) // 3
```

Elements can be inserted anywhere in the array, using the same zero-based index that we used in the preceding code:

```
gamesToPlay.insert("Breath of the Wild ", at: 2)
```

So, by inserting `"Breath of the Wild"` at index 2, it will be placed at the third position in our array, and all the elements at position 2 or greater will be moved down by one.

This increases the array's count:

```
print(gamesToPlay.count) // 4
```

The array also provides some helpful computed properties for accessing elements at either end of the array:

```
let firstGameToPlay = gamesToPlay.first ?? ""
print(firstGameToPlay) // "The Secret of Monkey Island"
let lastGameToPlay = gamesToPlay.last ?? ""
print(lastGameToPlay as Any) // "Alien Isolation"
```

These properties are optional values, as the array may be empty, and if it is, these will be `nil`. However, accessing an array element via an index subscript returns a non-optional value.

> **Note**
>
> In the preceding example, we used a nil-coalescing operator (`??`). This operator allows us to handle situations where the value is `nil` (e.g., if `gamesToPlay` was empty), but we need a default returned value (in this case, we return an empty string, `""`).

In addition to retrieving values via the subscript, we can also assign values to an array subscript:

```
gamesToPlay[1] = "Half Life 2 (2004)"
```

This will replace the element at the given index with the new value.

When we created our first array, we created an empty array and then appended values to it. Additionally, an array literal can be used to create an array that already contains values:

```
let graphicAdventureGames: [String] = ["Monkey Island 2",
    "Loom",
    "Sam & Max"]
```

An array type can be specified with the element type enclosed by square brackets, and the array literal can be defined by comma-separated elements within square brackets. So, we can define an array of integers like this:

```
let fibonacci: [Int] = [1, 1, 2, 3, 5, 8, 13, 21, 34, 55]
```

As we learned in *Chapter 1, Swift Fundamentals*, in the *Using the basic types – strings, ints, floats, and booleans* recipe, the compiler can often infer the type from the value we assign, and when the type is inferred, we don't need to specify it. In both the preceding arrays, `graphicAdventureGames` and `fibonacci`, all the elements in the array are of the same type – that is, `String` and `Int`, respectively. Since these types can be inferred, we don't need to define them:

```
let graphicAdventureGames = ["Monkey Island 2", "Loom", "Sam & Max"]

let fibonacci = [1, 1, 2, 3, 5, 8, 13, 21, 34, 55]
```

Arrays can be combined using the + operator:

```
gamesToPlay = gamesToPlay + graphicAdventureGames
```

This will create a new array by appending the elements in the second array to the first.

The array provides a convenience initializer that will fill an array with repeating elements. We can use this initializer to create an array with the name of a well-known movie trilogy:

```
var batmanGames = Array<String>(repeating: "Batman: ", count: 3)
```

We can then combine subscript access, string appending, and subscript assignment to add the full movie name to our trilogy array:

```
batmanGames[0] = batmanGames[0] + "Arkham Asylum"
batmanGames[1] = batmanGames[1] + "Arkham City"
batmanGames[2] = batmanGames[2] + "Arkham Knight"
```

The array also provides a helper to replace a range of values with the values contained in another array:

```
gamesToPlay.replaceSubrange(2...4, with: batmanGames)
```

Here, we have specified a range using . . . to indicate a range between two integer values, inclusive of those values. So, this range contains the 2, 3, and 4 integers.

We will specify ranges in this way in subsequent chapters. Alternatively, you can specify a range that goes up to, but not including, the top of the range. This is known as a half-open range:

```
gamesToPlay.replaceSubrange(2..<5, with: batmanGames)
```

For our arrays, we've added elements, accessed them, and replaced them, so we need to know how to remove elements from an array:

```
gamesToPlay.remove(at: 6)
```

Provide the index of the element to the `remove` method. By doing this, the element at that index will be removed from the array, and all the subsequent elements will move up one place to fill the empty space. This will reduce the array's count by 1:

```
print(gamesToPlay.count) // 6
```

There's more...

If you are familiar with Objective-C, you will have used `NSArray`, which provides similar functionalities to a Swift array. You may also remember that `NSArray` is immutable, which means its contents can't be changed once it's been created. If you need to change its contents, then `NSMutableArray` should be used instead. Due to this, you may be wondering if Swift has similar concepts of mutable and immutable arrays. It does, but rather than using separate mutable and immutable types, you create a mutable array by declaring it as a variable and an immutable array by declaring it as a constant:

```
let evenNumbersTo10 = [2, 4, 6, 8, 10] evenNumbersTo10.append(12) //
Doesn't compile

var evenNumbersTo12 = evenNumbersTo10 evenNumbersTo12.append(12) //
Does compile
```

To understand why this is the case, it's important to know that an array is a value type, as are the other collection types in Swift.

As we saw in *Chapter 1, Swift Fundamentals*, a value type is immutable in nature and creates a changed copy whenever it is mutated. Therefore, by assigning the array to a constant using `let`, we prevent any new value from being assigned, making mutating the array impossible.

See also

Further information about arrays can be found in Apple's documentation on the Swift language at `https://developer.apple.com/documentation/swift/array`.

Arrays use generics to define the element type they contain. Generics will be discussed in detail in *Chapter 4, Generics, Operators, and Nested Types*.

Further information about the nil-coalescing operator can be found at `https://docs.swift.org/swift-book/documentation/the-swift-programming-language/basicoperators/#Nil-Coalescing-Operator`

Containing your data in sets

The following collection type we will look at is a **set**. Sets differ from arrays in two important ways. The elements in a set are stored *unordered*, and each unique element is only held once. In this recipe, we will learn how to create and manipulate sets.

Getting ready

In this recipe, we can use the playground from the previous recipe. Don't worry if you didn't work through the previous recipe, as this one will contain all the code you need.

How to do it...

First, let's explore some ways we can create sets and perform set algebra on them:

1. Create an array that contains the first nine Fibonacci numbers, and also a set containing the same:

    ```
    let fibonacciArray: Array<Int> = [1, 1, 2, 3, 5, 8, 13, 21, 34]
    let fibonacciSet: Set<Int> = [1, 1, 2, 3, 5, 8, 13, 21, 34]
    ```

2. Print out the number of elements in each collection using the count property. Despite being created with the same elements, the count value is different:

```
print(fibonacciArray.count) // 9
print(fibonacciSet.count) // 8
```

3. Insert an element into a set of animals, remove an element, and check whether a set contains a given element:

```
var animals: Set<String> = ["cat", "dog", "mouse", "elephant"]
animals.insert("rabbit")
print(animals.contains("dog")) // true animals.remove("dog")
print(animals.contains("dog")) // false
```

4. Create some sets containing common mathematical number groups. We will use these to explore some methods for set algebra:

```
let evenNumbers = Set<Int>(arrayLiteral: 2, 4, 6, 8, 10)
let oddNumbers: Set<Int> = [1, 3, 5, 7, 9]
let squareNumbers: Set<Int> = [1, 4, 9]
let triangularNumbers: Set<Int> = [1, 3, 6, 10]
```

5. Obtain the union of two sets, and print the result:

```
let evenOrTriangularNumbers = evenNumbers.
union(triangularNumbers)
// 2, 4, 6, 8, 10, 1, 3, unordered
print(evenOrTriangularNumbers.count) // 7
```

6. Obtain the intersection of two sets, and print the result:

```
let oddAndSquareNumbers = oddNumbers.intersection(squareNumbers)
// 1, 9, unordered
print(oddAndSquareNumbers.count) // 2
```

7. Obtain the symmetric difference of two sets, and print the result:

```
let squareOrTriangularNotBoth = squareNumbers.
symmetricDifference(triangularNumbers)
// 4, 9, 3, 6, 10, unordered
print(squareOrTriangularNotBoth.count) // 5
```

8. Obtain the result of subtracting one set from another, and print the result:

```
let squareNotOdd = squareNumbers.subtracting(oddNumbers) // 4
print(squareNotOdd.count) // 1
```

Now, we will examine the set membership comparison methods that are available:

1. Create some sets with overlapping memberships:

    ```
    let animalKingdom: Set<String> = ["dog", "cat", "pigeon",
    "chimpanzee", "snake", "kangaroo", "giraffe", "elephant",
    "tiger", "lion", "panther"]
    let vertebrates: Set<String> = ["dog", "cat", "pigeon",
    "chimpanzee", "snake", "kangaroo", "giraffe", "elephant",
    "tiger", "lion", "panther"]
    let reptile: Set<String> = ["snake"]
    let mammals: Set<String> = ["dog", "cat", "chimpanzee",
    "kangaroo", "giraffe", "elephant", "tiger", "lion", "panther"]
    let catFamily: Set<String> = ["cat", "tiger", "lion", "panther"]
    let domesticAnimals: Set<String> = ["cat", "dog"]
    ```

2. Use the isSubset method to determine whether one set is a subset of another. Then, print the result:

    ```
    print(mammals.isSubset(of: animalKingdom)) // true
    ```

3. Use the isSuperset method to determine whether one set is a superset of another. Then, print the result:

    ```
    print(mammals.isSuperset(of: catFamily)) // true
    ```

4. Use the isStrictSubset method to determine whether one set is a strict subset of another. Then, print the result:

    ```
    print(vertebrates.isStrictSubset(of: animalKingdom)) // false
    print(mammals.isStrictSubset(of: animalKingdom)) // true
    ```

5. Use the isStrictSuperset method to determine whether one set is a strict superset of another. Then, print the result:

    ```
    print(animalKingdom.isStrictSuperset(of: vertebrates)) // false
    print(animalKingdom.isStrictSuperset(of: domesticAnimals))// true
    ```

6. Use the isDisjoint method to determine whether one set is disjointed with another. Then, print the result:

    ```
    print(catFamily.isDisjoint(with: reptile)) // true
    ```

How it works...

Sets are created in almost the same way as arrays, and like arrays, we have to specify the element type that we will be stored in them:

```
let fibonacciArray: Array<Int> = [1, 1, 2, 3, 5, 8, 13, 21, 34]
let fibonacciSet: Set<Int> = [1, 1, 2, 3, 5, 8, 13, 21, 34]
```

Arrays and sets store their elements differently. If you provide multiple elements of the same value to an array, it will store them multiple times. A set works differently; it will only store one version of each unique element. Therefore, in the preceding Fibonacci number sequence, the array stores two elements for the first two values, `1, 1`, but the set will store this as just one `1` element. This leads to the collections having different counts, despite being created with the same values:

```
print(fibonacciArray.count) // 9
print(fibonacciSet.count) // 8
```

This ability to store elements uniquely is made possible due to a requirement that a set has, regarding the type of elements it can hold. A set's elements must conform to the `Hashable` protocol. This protocol requires a `hashValue` property to be provided as `Int`, and the set uses this `hashValue` to do its uniqueness comparison. Both the `Int` and `String` types conform to `Hashable`, but any custom types that will be stored in a set will also need to conform to `Hashable`.

A set's `insert`, `remove`, and `contains` methods work as you would expect, with the compiler enforcing that the correct types are provided. This compiler type checking is done thanks to the `generics` constraints that all the collection types have. We will cover generics in more detail in *Chapter 4, Generics, Operators, and Nested Types*.

Union

The `union` method returns a set containing all the unique elements from the set that the method is called on, as well as the set that was provided as a parameter:

```
let evenOrTriangularNumbers = evenNumbers.union(triangularNumbers)
// 2,4,6,8,10,1,3,unordered
```

The following diagram depicts the union of **Set A** and **Set B**:

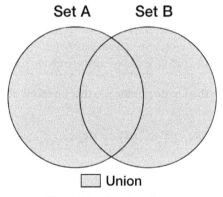

Figure 2.2 – A union of sets

Intersection

The `intersection` method returns a set of unique elements that were contained in both the set that the method was called on and the set that was provided as a parameter:

```
let oddAndSquareNumbers = oddNumbers.intersection(squareNumbers)
// 1, 9, unordered
```

The following diagram depicts the intersection of **Set A** and **Set B**:

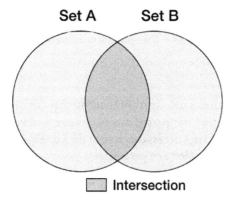

Figure 2.3 – The set intersection

Symmetric difference

The `symmetricDifference` method returns a set of unique elements that are in either the set the method is called on or the set that's provided as a parameter, but not elements that are in both:

```
let squareOrTriangularNotBoth = squareNumbers.
symmetricDifference(triangularNumbers)
// 4, 9, 3, 6, 10, unordered
```

> **Note**
>
> This set operation is sometimes referred to as an `exclusiveOr` method by other programming languages, including previous versions of Swift.

The following diagram depicts the symmetric difference between **Set A** and **Set B**:

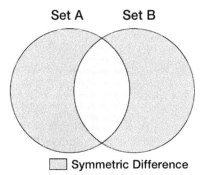

Figure 2.4 – The symmetric difference

Subtracting

The subtracting method returns a unique set of elements that can be found in the set the method was called on, but not in the set that was passed as a parameter. Unlike the other set manipulation methods we've mentioned, this will not necessarily return the same value if you swap the set that the method is called on with the set provided as a parameter:

```
let squareNotOdd = squareNumbers.subtracting(oddNumbers) // 4
```

The following diagram depicts the set that's created by subtracting **Set B** from **Set A**:

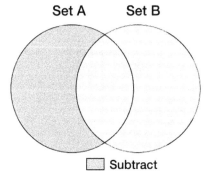

Figure 2.5 – Subtracting a set

A membership comparison

In addition to set manipulation methods, there are a number of methods we can use to determine information about set membership.

The isSubset method will return true if all the elements in the set that the method is called on are contained within the set that's passed as a parameter:

```
print(mammals.isSubset(of: animalKingdom)) // true
```

The following diagram depicts **Set B** as the subset of **Set A**:

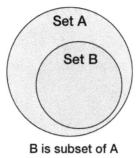

B is subset of A

Figure 2.6 – The subset

This will also return `true` if the two sets are equal (i.e., they contain the same elements). If you only want a `true` value when the set that the method is called on is a subset and not equal, then you can use `isStrictSubset`:

```
print(vertebrates.isStrictSubset(of: animalKingdom)) // false
print(mammals.isStrictSubset(of: animalKingdom)) // true
```

The `isSuperset` method will return `true` if all the elements in the set that have been passed as a parameter are within the set that the method is called on:

```
print(mammals.isSuperset(of: catFamily)) // true
```

The following diagram depicts **Set A** as the superset of **Set B**:

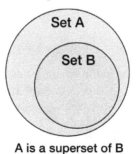

A is a superset of B

Figure 2.7 – The superset

This will also return `true` if the two sets are equal (i.e., they contain the same elements). If you only want a `true` value when the set that the method is called on is a superset and not equal, then you can use `isStrictSuperset`:

```
print(animalKingdom.isStrictSuperset(of: vertebrates)) // false
print(animalKingdom.isStrictSuperset(of: domesticAnimals)) // true
```

The isDisjoint method will return true if there are no common elements between the set that the method is called on and the set that was passed as a parameter:

```
print(catFamily.isDisjoint(with: reptile)) // true
```

The following diagram shows that **Set A** and **Set B** are disjointed:

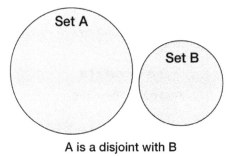

A is a disjoint with B

Figure 2.8 – A disjoint

As with arrays, a set can be declared immutable by assigning it to a let constant instead of a var variable:

```
let planets: Set<String> = ["Mercury", "Venus", "Earth", "Mars",
"Jupiter", "Saturn", "Uranus", "Neptune", "Pluto"]
planets.remove("Pluto") // Doesn't compile
```

This declaration is possible because a set, like the other collection types, is a value type. Removing an element would mutate the set, which creates a new copy, but a let constant can't have a new value assigned to it, so the compiler prevents any mutating operations.

See also

Further information about arrays can be found in Apple's documentation on the Swift language at https://docs.swift.org/swift-book/documentation/the-swift-programming-language/collectiontypes/.

Sets use generics to define the element types they contain. Generics will be discussed in detail in *Chapter 4, Generics, Operators, and Nested Types*.

Storing key-value pairs with dictionaries

The last collection type we will look at is the **dictionary**. This is a familiar construct in programming languages, where it is sometimes referred to as a *hash table*. A dictionary holds a collection of pairings between a key and a value. The **key** can be any element that conforms to the Hashable protocol (just like elements in a set), while the **value** can be any type. The contents of a dictionary are not stored in order, unlike an array; instead, key is used both when storing a value and as a lookup when retrieving a value.

Getting ready

In this recipe, we will use a dictionary to store details of people at a place of work. We need to store and retrieve a person's information based on their role in an organization, such as a company directory. To hold this person's information, we will use a modified version of our `Person` class from *Chapter 1, Swift Fundamentals*.

Enter the following code into a new playground:

```
struct PersonName {
  let givenName: String
  let familyName: String
}

enum CommunicationMethod {
  case phone
  case email
  case textMessage
  case fax
  case telepathy
  case subSpaceRelay
  case tachyons
}

class Person {
    let name: PersonName
    let preferredCommunicationMethod: CommunicationMethod

    convenience init(givenName: String, familyName: String,
commsMethod: CommunicationMethod) {
        let name = PersonName(givenName: givenName, familyName:
familyName)
        self.init(name: name, commsMethod: commsMethod)
    }

    init(name: PersonName, commsMethod: CommunicationMethod) {
        self.name = name
        preferredCommunicationMethod = commsMethod
    }

    var displayName: String {
        return "\(name.givenName) \(name.familyName)"
    }
}
```

How to do it...

Let's use the `Person` object we defined previously to build up our workplace directory using a dictionary:

1. Create a Dictionary for the employee directory:

```
var crew = Dictionary<String, Person>()
```

2. Populate the dictionary with employee details:

```
crew["Captain"] = Person(givenName: "Jean-Luc", familyName:
"Picard", commsMethod: .phone)
crew["First Officer"] = Person(givenName: "William", familyName:
"Riker", commsMethod: .email)
crew["Chief Engineer"] = Person(givenName: "Geordi", familyName:
"LaForge", commsMethod: .textMessage)
crew["Second Officer"] = Person(givenName: "Data", familyName:
"Soong", commsMethod: .fax)
crew["Councillor"] = Person(givenName: "Deanna", familyName:
"Troi", commsMethod: .telepathy)
crew["Security Officer"] = Person(givenName: "Tasha",
familyName: "Yar", commsMethod: .subSpaceRelay)
crew["Chief Medical Officer"] = Person(givenName: "Beverly",
familyName: "Crusher", commsMethod: .tachyons)
```

3. Retrieve an array of all the keys in the dictionary. This will give us an array of all the roles in the organization:

```
let roles = Array(crew.keys)
print(roles)
```

4. Use a key to retrieve one of the employees, and print the result:

```
let firstRole = roles.first! // Chief Medical Officer
let cmo = crew[firstRole]! // Person: Beverly Crusher
print("\(firstRole): \(cmo.displayName)") // Chief Medical
Officer: Beverly Crusher
```

5. Replace a value in the dictionary by assigning a new value against an existing key. The previous value for the key is discarded when a new value is set:

```
print(crew["Security Officer"]!.name.givenName) // Tasha
crew["Security Officer"] = Person(givenName: "Worf", familyName:
"Son of Mogh", commsMethod: .subSpaceRelay)
print(crew["Security Officer"]!.name.givenName) // Worf
```

With that, we have learned how to create, populate, and look up values in a dictionary.

How it works...

As with the other collection types, when we create a dictionary, we need to provide the types that the dictionary will hold. For dictionaries, there are two types that we need to define. The first is the type of the key (which must conform to `Hashable`), while the second is the type of the value being stored against the key. For our dictionary, we are using `String` for the key and `Person` for the values being stored:

```
var crew = Dictionary<String, Person>()
```

As with an array, we can specify a `dictionary` type using square brackets and create one using a dictionary literal, where `:` separates the key and the value:

```
let intByName: [String: Int] = ["one": 1, "two": 2, "three": 3]
```

Therefore, we can change our dictionary definition so that it looks like this:

```
var crew: [String: Person] = [:]
```

The `[:]` symbol denotes an empty dictionary as a dictionary literal.

Elements are added to a dictionary using a subscript. Unlike an array, which takes an `Int` index in the subscript, a dictionary takes the key and then pairs the given value with the given key. In the following example, we assign a `Person` object to the `"Captain"` key:

```
crew["Captain"] = Person(givenName: "Jean-Luc",
familyName: "Picard", commsMethod: .phone)
```

If no value currently exists, the assigned value will be added. If a value already exists for the given key, the old value will be replaced with the new value, and the old value will be discarded.

There are properties in the dictionary that provide all the keys and values. These properties are of a custom collection type, which can be passed to an array initializer to create an array:

```
let roles = Array(crew.keys) print(roles)
```

To display all the dictionary's keys, as provided by the `keys` property, we can either create an array or iterate over the collection directly. We will cover iterating over a collection's values in *Chapter 3, Data Wrangling with Swift*, so for now, we will create an array.

Now, we will use one of the values from an array of keys, alongside the crew, to retrieve full details about the associated `Person`:

```
let firstRole = roles.first! // Chief Medical Officer
let cmo = crew[firstRole]! // Person: Beverly Crusher
print("\(firstRole): \(cmo.displayName)") // Chief Medical Officer:
Beverly Crusher
```

We get the first element using the `first` property, but since this is an optional type, we need to force-unwrap it using `!`. We can pass `firstRole`, which is now a non-optional `String` to the dictionary subscript, to get the `Person` object associated with that key. The return type to retrieve the value via subscript is also optional, so it also needs to be **force-unwrapped** before we print its values.

> **Note**
>
> Force unwrapping is usually an unsafe thing to do, as if we force unwrap a value that turns out to be nil, our code will crash. We advise you to check that a value isn't nil before unwrapping the optional. We will cover how to do this in *Chapter 3*.

There's more...

In this recipe, we used strings as the keys for our dictionary. However, we can also use a type that conforms to the `Hashable` protocol.

One downside of using `String` as a key for our employee directory is that it is very easy to mistype an employee's role or look for a role that you expect to exist but doesn't. So, we can improve our implementation by using something that conforms to `Hashable` and is better suited to being used as a key in our model.

We have a finite set of employee roles in our model, and an enumeration is perfect for representing a finite number of options, so let's define our roles as an enum:

```
enum Role: String {
  case captain = "Captain"
  case firstOfficer = "First Officer"
  case secondOfficer = "Second Officer"
  case chiefEngineer = "Chief Engineer"
  case councillor = "Councillor"
  case securityOfficer = "Security Officer"
  case chiefMedicalOfficer = "Chief Medical Officer"
}
```

Now, let's change our `Dictionary` definition so that it uses this new enum as a key, and then insert our employees using these enum values:

```
var crew = Dictionary<Role, Person>()
crew[.captain] = Person(givenName: "Jean-Luc", familyName: "Picard",
commsMethod: .phone)
crew[.firstOfficer] = Person(givenName: "William", familyName:
"Riker", commsMethod: .email)
crew[.chiefEngineer] = Person(givenName: "Geordi", familyName:
"LaForge", commsMethod: .textMessage)
crew[.secondOfficer] = Person(givenName: "Data", familyName: "Soong",
```

```
commsMethod: .fax)
crew[.councillor] = Person(givenName: "Deanna", familyName: "Troi",
commsMethod: .telepathy)
crew[.securityOfficer] = Person(givenName: "Tasha", familyName: "Yar",
commsMethod: .subSpaceRelay)
crew[.chiefMedicalOfficer] = Person(givenName: "Beverly", familyName:
"Crusher", commsMethod: .tachyons)
```

You will also need to change all the other uses of `crew` so that they use the new enum-based key.

Let's take a look at how and why this works. We created `Role` as a `String`-based enum:

```
enum Role: String {
    //...
}
```

Defining it in this way has two benefits:

- We intend to display these roles to the user, so we will need a string representation of the `Role` enum, regardless of how we defined it.

- Enums have a little bit of protocol and generics magic in them, which means that if an enum is backed by a type that implements the `Hashable` protocol (as `String` does), the enum also automatically implements the `Hashable` protocol. Therefore, defining `Role` as being `String`-based satisfies the dictionary requirement of a key being `Hashable`, without us having to do any extra work.

With our `crew` dictionary now defined as having a `Role`-based key, all subscript operations have to use a value in the enum role:

```
crew[.captain] = Person(givenName: "Jean-Luc", familyName: "Picard",
commsMethod: .phone)
let cmo = crew[.chiefMedicalOfficer]
```

The compiler enforces this, so it's no longer possible to use an incorrect role when interacting with our employee directory. This pattern of using Swift's constructs and type system to enforce the correct use of code is something we should strive to do, as it can reduce bugs and prevent our code from being used in unexpected ways.

See also

Further information about dictionaries can be found in Apple's documentation on the Swift language at `http://swiftbook.link/docs/collections`.

Subscripts for custom types

By using collection types, we have seen that their elements are accessed through subscripts. However, it's not just collection types that can have subscripts; your own custom types can provide subscript functionality too.

Getting ready

In this recipe, we will create a simple game of *tic-tac-toe*, also known as *noughts and crosses*. To do this, we need a three-by-three grid of positions, with each position being filled by either a nought from player 1, a cross from player 2, or nothing. We can store these positions in an array of arrays.

The initial game setup code uses the concepts we've already covered in this book, so we won't go into its implementation. Enter the following code into a new playground so that we can see how subscripts improve its usage:

```
enum GridPosition: String {
 case player1 = "o"
 case player2 = "x"
 case empty = " "
}

struct TicTacToe {
 var gridStorage: [[GridPosition]] = []
 init() {
  gridStorage.append(Array(repeating: .empty,
    count: 3))
  gridStorage.append(Array(repeating: .empty,
    count: 3))
  gridStorage.append(Array(repeating: .empty,
    count: 3))
 }
 func gameStateString() -> String {
  var stateString = "\n"
  stateString += printableString(forRow: gridStorage[0])
  stateString += "\n"
  stateString += printableString(forRow: gridStorage[1])
  stateString += "\n"
  stateString += printableString(forRow: gridStorage[2])
  stateString += "\n"
  return stateString
 }
 func printableString(forRow row: [GridPosition]) -> String {
  var rowString = "| \(row[0].rawValue) "
```

```
rowString += "| \(row[1].rawValue) "
rowString += "| \(row[2].rawValue) |\n"
return rowString
  }
}
```

How to do it...

Let's run through how we can use the tic-tac-toe game defined previously, as well as how we can improve how it is used, using a subscript. We will also examine how this works:

1.   Let's create an instance of our `TicTacToe` grid:

```
var game = TicTacToe()
```

2.   For a player to make a move, we need to change the `GridPosition` value that's been assigned to the relevant place in the array of arrays. This is used to store the grid positions. Player 1 will place a nought in the middle position of the grid, which would be row position *1* and column position *1* (since it's a zero-based array):

```
// Move 1
game.gridStorage[1][1] = .player1
print(game.gameStateString())
/*
```

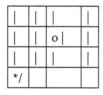

Figure 2.9 – The grid positions

3.   Then, player 2 places their cross in the top-right position, which is row position *0* and column position *2*:

```
// Move 2
game.gridStorage[0][2] = .player2
print(game.gameStateString())
/*

||| x |

|| o ||
```

```
||||

*/
```

We can make moves in our game. We can do this by adding information directly to the `gridStorage` array, which isn't ideal. The player shouldn't need to know how the moves are stored, and we should be able to change how we store the game information without having to change how the moves are made. To solve this, let's create a subscript of our game struct so that making a move in the game is just like assigning a value to an array.

4. Add the following `subscript` method to the `TicTacToe` struct:

```
struct TicTacToe {
 var gridStorage: [[GridPosition]] = []
 //...
 subscript(row: Int, column: Int) -> GridPosition {
   get {
     return gridStorage[row][column]
   }
   set(newValue) {
     gridStorage[row][column] = newValue
   }
 }
 //...
}
```

5. So, now, we can change how each player makes their move and finish the game:

```
// Move 1
game[1, 1] = .player1 print(game.gameStateString())
/*
```

```
|  |  |  |  |  |
|  |  | o|  |  |
|  |  |  |  |  |
*/
```

```
// Move 2
game[0, 2] = .player2
print(game.gameStateString())
/*

||| x |

|| o ||
```

```
| | | |

*/

// Move 3
game[0, 0] = .player1
print(game.gameStateString())
/*

| o || x |

|| o ||

| | | |

*/

// Move 4
game[1, 2] = .player2
print(game.gameStateString())
/*

| o || x |

|| o | x |

| | | |

*/

// Move 5
game[2, 2] = .player1
print(game.gameStateString())
/*

| o || x |

|| o | x |

||| o |

*/
```

6. Just like when using an array, we can use a subscript to access the value, as well as assign a value
    to it:

```
let topLeft = game[0, 0]
let middle = game[1, 1]
let bottomRight = game[2, 2]
let p1HasWon = (topLeft == .player1) && (middle == .player1) &&
(bottomRight == .player1)
```

How it works...

Subscript functionality can be defined within a class, struct, or enum, or declared within a protocol
as a requirement. To do this, we can define `subscript` (which is a reserved keyword that activates
the required functionality) with input parameters and an output type:

```
subscript(row: Int, column: Int) -> GridPosition
```

This `subscript` definition works like a computed property, where `get` can be defined to allow you
to access values through `subscript`, and `set` can be defined to assign values using `subscript`:

```
subscript(row: Int, column: Int) -> GridPosition {
  get {
    return gridStorage[row][column]
  }
  set(newValue) {
    gridStorage[row][column] = newValue
  }
}
```

Any number of input parameters can be defined, and these should be provided as comma-separated
values in the subscript:

```
game[1, 2] = .player2 // Assigning a value
let topLeft = game[0, 0] // Accessing a value
```

There's more...

Just like parameters defined in a function, `subscript` parameters can have additional labels. If defined,
these become required at the call site, so the subscript we added can alternatively be defined as follows:

```
subscript(atRow row: Int, atColumn column: Int) -> GridPosition
```

In this case, when using `subscript`, we would also provide the labels in it:

```
game[atRow: 1, atColumn: 2] = .player2 // Assigning a value
let topLeft = game[atRow: 0, atColumn: 0] // Accessing a value
```

See also

Further information about subscripts can be found in Apple's documentation on the Swift language at `http://swiftbook.link/docs/subscripts`.

Changing your name with a type alias

The `typealias` declaration allows you to create an alias for a type (and is, therefore, pretty accurately named!). You can specify a name that can be used in place of any given type of definition. If this type is quite complex, a type alias can be a useful way to simplify its use.

Getting ready

In this recipe, we won't be using any components from the previous recipes, so you can create a new playground for this recipe.

How to do it...

We will use a type alias to replace an array definition:

1. First, let's create something we can store in an array. In this instance, let's create a Pug struct:

    ```
    struct Pug {
      let name: String
    }
    ```

2. Now, we can create an array that will contain instances of a Pug struct:

    ```
    let pugs = [Pug]()
    ```

> **Note**
>
> You may or may not know that the collective noun for a group of pugs is called `grumble`.

3. We can set up `typealias` to define an array of pugs as `Grumble`:

    ```
    typealias Grumble = [Pug]
    ```

4. With this defined, we can substitute `Grumble` wherever we would use `[Pug]` or `Array<Pug>`:

    ```
    var grumble = Grumble()
    ```

5. However, this isn't some new type – it is just an array with all the same functionalities:

    ```
    let marty = Pug(name: "Marty McPug")
    let wolfie = Pug(name: "Wolfgang Pug")
    ```

```
let buddy = Pug(name: "Buddy")
grumble.append(marty)
grumble.append(wolfie)
grumble.append(buddy)
```

There's more...

The preceding example allows us to use types in a more natural and expressive way. In addition, we can use `typealias` to simplify a more complex type that may be used in multiple places.

To see how this might be useful, we can partially build an object to fetch program information:

```
enum Channel {
 case BBC1
 case BBC2
 case BBCNews
 //...
}

class ProgrammeFetcher {
 func fetchCurrentProgrammeName(forChannel channel: Channel,
resultHandler: (String?, Error?) -> Void) {
  // ...
  // Do the work to get the current programme
  // ...
  let exampleProgramName = "Sherlock"
  resultHandler(exampleProgramName, nil)

 }
 func fetchNextProgrammeName(forChannel channel: Channel,
resultHandler: (String?, Error?) -> Void) {
  // ...
  // Do the work to get the next programme
  // …
  let exampleProgramName = "Luther"
  resultHandler(exampleProgramName, nil)
 }
}
```

In the `ProgrammeFetcher` object, we have two methods that take a channel and a result handler closure. The result handler closure has the following definition. We have to define this twice, once for each method:

```
(String?, Error?) -> Void
```

Alternatively, we can define this closure definition with a typealias called `FetchResultHandler` and replace each method definition with a reference to this `typealias`:

```
class ProgrammeFetcher {
 typealias FetchResultHandler = (String?, Error?) -> Void
 func fetchCurrentProgrammeName(forChannel channel: Channel,
resultHandler: FetchResultHandler) {
  // Get next programme
  let programmeName = "Sherlock"
  resultHandler(programmeName, nil)
 }


 func fetchNextProgrammeName(forChannel channel: Channel,
resultHandler: FetchResultHandler) {
  // Get next programme
  let programmeName = "Luther"
  resultHandler(programmeName, nil)
 }
}
```

Not only does this save us from defining the closure type twice, but it is also a better description of the function that the closure performs.

Using `typealias` also doesn't affect how we provide closure to the method:

```
let fetcher = ProgrammeFetcher()
fetcher.fetchCurrentProgrammeName(forChannel: .BBC1,
resultHandler: { programmeName, error in
 print(programmeName as Any)
})
```

See also

Further information about type alias can be found in Apple's documentation on the Swift language at `https://docs.swift.org/swift-book/documentation/the-swift-programming-language/declarations/`.

Getting property changing notifications using property observers

It's common to want to know when a property's value changes. Perhaps you want to update the value of another property or update some UI element. In Objective-C, this was often accomplished by writing your own getter and setter or using **Key-Value Observing (KVO)**. However, in Swift, we have native support for property observers.

Getting ready

To examine property observers, we should create an object with a property that we want to observe. Let's create an object that manages users and a property that holds the current user's name.

Enter the following code into a new playground:

```
class UserManager {
  var currentUserName: String = "Emmanuel Goldstein"
}
```

We want to present some friendly messages when the current user changes. We'll use property observers to do this.

How to do it...

Let's get started:

1. Amend the currentUserName property definition so that it looks as follows:

    ```
    class UserManager {
      var currentUserName: String = "Guybrush Threepwood" {
        willSet (newUserName) {
          print("Goodbye to \(currentUserName)")
          print("I hear \(newUserName) is on their way!")
        }
        didSet (oldUserName) {
          print("Welcome to \(currentUserName)")
          print("I miss \(oldUserName) already!")
        }
      }
    }
    ```

2. Create an instance of UserManager, and change the current username. This will generate friendly messages:

    ```
    let manager = UserManager()
    manager.currentUserName = "Elaine Marley"
    // Goodbye to Guybrush Threepwood
    // I hear Elaine Marley is on their way!
    // Welcome to Elaine Marley
    // I miss Guybrush Threepwood already!

    manager.currentUserName = "Ghost Pirare LeChuck"
    ```

```
// Goodbye to Elaine Marley
// I hear Ghost Pirare LeChuck is on their way!
// Welcome to Ghost Pirare LeChuck
// I miss Elaine Marley already!
```

How it works...

Property observers can be added within curly brackets after the property declaration, and there are two types – willSet and didSet.

The willSet observer will be called before the property is set and provides the value that will be set on the property. This new value can be given a name within brackets – for example, newUserName:

```
willSet (newUserName) {
 //...
}
```

The didSet observer will be called after the property is set and provides the value that the property had before being set. This old value can be given a name within brackets – for example, oldUserName:

```
didSet (oldUserName) {
 //...
}
```

There's more...

The new value and old value that are passed into the property observers have implicit names, so there is no need to explicitly name them. The willSet observer is passed a value with an implicit name of newValue, and the didSet observer is passed a value with an implicit name of oldValue.

Therefore, we can remove our explicit names and use the implicit value names:

```
class UserManager {
  var currentUserName: String = "Guybrush Threepwood" {
    willSet {
      print("Goodbye to \(currentUserName)")
      print("I hear \(newValue) is on their way!")
    }
    didSet {
      print("Welcome to \(currentUserName)")
      print("I miss \(oldValue) already!")
    }
  }
}
```

See also

Further information about property observers can be found in Apple's documentation on the Swift language at `http://swiftbook.link/docs/properties`.

Extending functionality with extensions

Extensions let us add functionality to our existing classes, structs, enums, and protocols. These can be especially useful when the original type is provided by an external framework, which means you aren't able to add functionality directly.

Imagine that we often need to obtain the first word from a given string. Rather than repeatedly writing the code to split the string into words and then retrieving the first word, we can extend the functionality of `String` to provide its own first word.

Getting ready

In this recipe, we won't be using any components from the previous recipes, so you can create a new playground for this recipe.

How to do it...

Let's get started:

1. Create an extension of `String`:

    ```
    extension String {

    }
    ```

2. Within the extension's curly brackets, add a function that returns the first word from the string:

    ```
    func firstWord() -> String {
     let spaceIndex = firstIndex(of: " ") ?? endIndex
     let word = prefix(upTo: spaceIndex)
     return String(word)
    }
    ```

3. Now, we can use this new method on `String` to get the first word from a phrase:

    ```
    let llap = "Ask me about Loom"
    let firstWord = llap.firstWord()
    print(firstWord) // Ask
    ```

How it works...

We can define an extension using the `extension` keyword and then specify the type we want to extend. The implementation of this extension is defined within curly brackets:

```
extension String {
  //...
}
```

Methods and computed properties can be defined in extensions in the same way that they can be defined within classes, structs, and enums. Here, we will add a `firstWord` function to the `String` struct:

```
extension String {
  func firstWord() -> String {
    let spaceIndex = firstIndex(of: " ") ?? endIndex
    let word = prefix(upTo: spaceIndex)
    return String(word)
  }
}
```

The implementation of the `firstWord` method is not important for this recipe, so we'll just touch on it briefly.

In Swift, `String` is a collection, so we can use the collection methods to find the first index of an empty space. However, this could be `nil`, since the string may contain only one word or no characters at all, so if the index is `nil`, we must use `endIndex` instead. The nil coalescing operator (`??`) is only used to assign `endIndex` if `firstIndex(of: " ")` is `nil`.

More generally, the operation will evaluate the value on the left-hand side of the operator, unless it is `nil`, in which case it will assign the value on the right-hand side.

Then, we use the index of the first space to retrieve the substring up to the index, which has a `SubString` type. We then use that to create and return `String`.

Extensions can implement anything that uses the existing functionality, but they can't store information in a new property. Therefore, computed properties can be added, but stored properties cannot. Let's change our `firstWord` method so that it's a computed property instead:

```
extension String {
  var firstWord: String {
    let spaceIndex = firstIndex(of: " ") ?? endIndex
    let word = prefix(upTo: spaceIndex)
    return String(word)
  }
}
```

There's more...

Extensions can also be used to add protocol conformance, so let's create a protocol that we want to add conformance to:

1. The protocol declares that something can be represented as `Int`:

```
protocol IntRepresentable {
 var intValue: Int { get }
}
```

2. We can extend `Int` and have it conform to `IntRepresentable` by returning itself:

```
extension Int: IntRepresentable {
 var intValue: Int {
  return self
 }
}
```

3. Now, we'll extend `String`, and we'll use an `Int` constructor that takes `String` and returns `Int` if our `String` contains digits that represent an integer:

```
extension String: IntRepresentable {
 var intValue: Int {
  return Int(self) ?? 0
 }
}
```

4. We can also extend our own custom types and add conformance to the same protocol, so let's create an enum that can be `IntRepresentable`:

```
enum CrewComplement: Int {
 case enterpriseD = 1014
 case voyager = 150
 case deepSpaceNine = 2000
}
```

5. Since our enum is `Int`-based, we can conform to `IntRepresentable` by providing `rawValue`:

```
extension CrewComplement: IntRepresentable {
 var intValue: Int {
  return rawValue
 }
}
```

6. We now have `String`, `Int`, and `CrewComplement` all conforming to `IntRepresentable`, and since we didn't define `String` or `Int`, we have only been able to add conformance through the use of extensions. This common conformance allows us to treat them as the same type:

```
var intableThings = [IntRepresentable] ()
intableThings.append(55)
intableThings.append(1200)
intableThings.append("5")
intableThings.append("1009")
intableThings.append(CrewComplement.enterpriseD)
intableThings.append(CrewComplement.voyager)
intableThings.append(CrewComplement.deepSpaceNine)
let over1000 = intableThings.compactMap {
  $0.intValue > 1000 ? $0.intValue: nil }
print(over1000)
```

The preceding example includes the use of `compactMap` and the ternary operator (`?`), which haven't been covered in this book. Further information can be found in the *See also* section.

See also

Further information about extensions can be found in Apple's documentation on the Swift language at `http://swiftbook.link/docs/extensions`.

The documentation for `compactMap` can be found at `https://developer.apple.com/documentation/swift/sequence/compactmap(_:)`.

Further information about the ternary operator can be found at `https://docs.swift.org/swift-book/documentation/the-swift-programming-language/basicoperators/#Ternary-Conditional-Operator`.

Controlling access with access control

Swift provides fine-grained access control, allowing you to specify the visibility that your code has to other areas of code. This enables you to be deliberate about the interface you provide to other parts of the system, thus encapsulating implementation logic and helping separate the areas of concern.

Swift has five access levels:

- **Private**: Only accessible within the existing scope (defined by curly brackets) or extensions in the same file
- **File private**: Accessible to anything in the same file, but nothing outside the file
- **Internal**: Accessible to anything in the same module, but nothing outside the module

- **Public**: Accessible both inside and outside the module, but cannot be subclassed or overwritten outside of the defining module

- **Open**: Accessible everywhere, with no restrictions in terms of its use, and can therefore be subclassed and overwritten

These can be applied to types, properties, and functions.

Getting ready

To explore each of these access levels, we need to step outside our playground comfort zone and create a module. To have something that will hold our module and a playground that can use it, we will need to create an *Xcode* workspace:

1. In *Xcode*, select **File** | **New** | **Workspace...** from the menu:

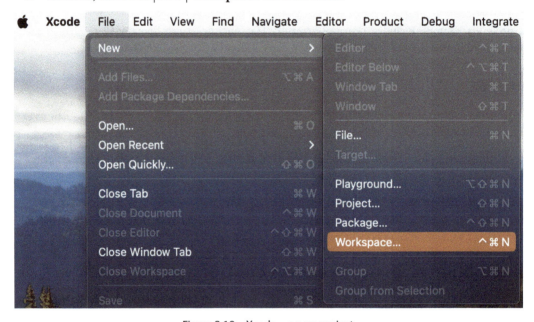

Figure 2.10 – Xcode – a new project

2. Give your workspace a name, such as `AccessControl`, and choose a save location. You will now see an empty workspace:

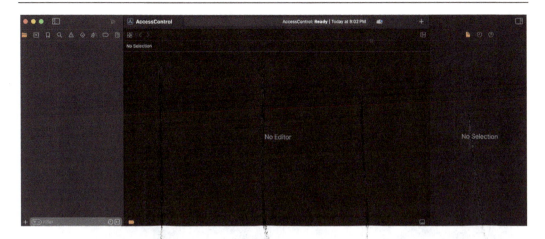

Figure 2.11 – Xcode – a new project structure

In this workspace, we need to create a module. To illustrate the access controls that are available, let's have our module represent something that tightly controls which information it exposes, and which information it keeps hidden. One thing that fits this definition is *Apple* – that is, the company.

3. Create a new project from the Xcode menu by selecting **File** | **New** | **Project...**:

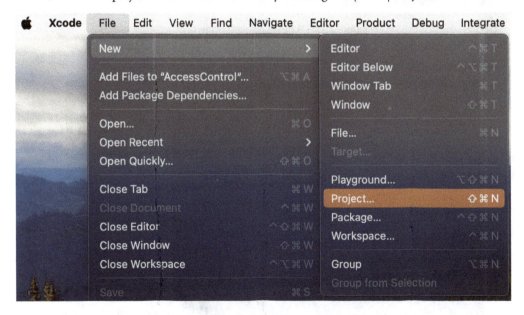

Figure 2.12 – A new project

4. From the template selector, select **Framework**:

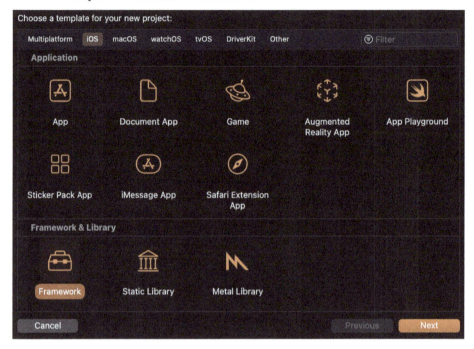

Figure 2.13 – A new project framework

5. Name the project `AppleInc`:

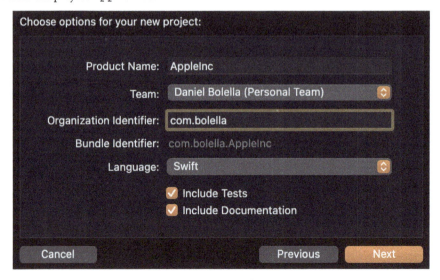

Figure 2.14 – Naming the project

6. Choose a location. Then, at the bottom of the window, ensure that **Add to:** has been set to the workspace we just created:

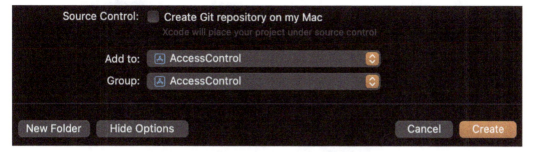

Figure 2.15 – The new project workspace group

7. Now that we have a module, let's set up a playground to use it in. From the Xcode menu, select **File | New | Playground…**:

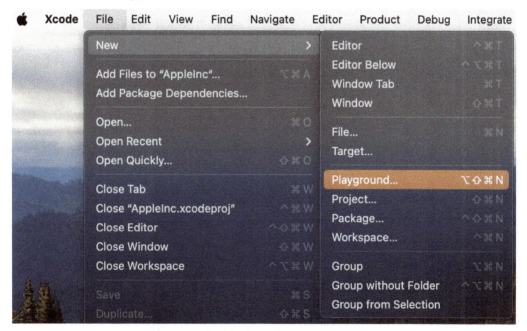

Figure 2.16 – A new playground

8. Give the playground a name, and add it to your workspace:

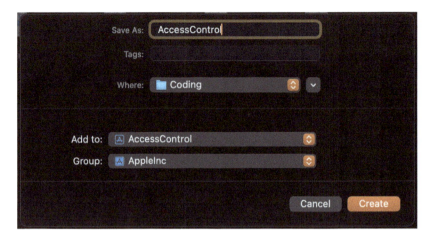

Figure 2.17 – A new project

9. Press the *run* button on the Xcode toolbar to build the **AppleInc** module:

Figure 2.18 – The Xcode toolbar

10. Select the playground from the file navigator, and add an import statement to the top of the file:

```
import AppleInc
```

We are now ready to look into the different access controls that are available.

How to do it...

Let's investigate the most restrictive of the access controls – `private`. Structures marked as `private` are only visible within the scope of the type they have been defined in, as well as any extensions of that type that are located in the same file. We know that *Apple* has super-secret areas where it works on its new products, so let's create one:

1. Select the `AppleInc` group in the file navigator, and create a new file by selecting **File** | **New** | **File...** from the menu. Let's call it `SecretProductDepartment`.

2. In this new file, create a `SecretProductDepartment` class using the `private` access control:

```
class SecretProductDepartment {
  private var secretCodeWord = "Titan"
  private var secretProducts = ["Apple Glasses", "Apple Car",
```

```
"Apple Brain Implant"]

func nextProduct(codeWord: String) -> String? {
  let codeCorrect = codeWord == secretCodeWord
  return codeCorrect ? secretProducts.first : nil
  }
}
```

3. Now, let's look at the `fileprivate` access control. Structures marked as `fileprivate` are only visible within the file that they are defined in, so a collection of related structures defined in the same file will be visible to each other, but anything outside the file will not see these structures.

    When you buy an iPhone from the Apple Store, it's not made in-store; it's made in a factory that the public doesn't have access to. So, let's model this using `fileprivate`.

    Create a new file called `AppleStore`. Then, create structures for `AppleStore` and `Factory` using the `fileprivate` access control:

```
public enum DeviceModel {
  case iPhone13
  case iPhone13Mini
  case iPhone13Pro
  case iPhone13ProMax
  }

public class AppleiPhone {
  public let model: DeviceModel
  fileprivate init(model: DeviceModel) {
    self.model = model
  }
  }

fileprivate class Factory {
  func makeiPhone(ofModel model: DeviceModel) -> AppleiPhone {
    return AppleiPhone(model: model)
  }
  }

public class AppleStore {
  private var factory = Factory()
  public func selliPhone(ofModel model: DeviceModel)
  -> AppleiPhone {
    return factory.makeiPhone(ofModel: model)
  }
  }
```

To investigate the `public` access control, we will define something that is visible outside the defining module but cannot be subclassed or overridden.

Apple itself is the perfect candidate to model this behavior, as certain parts of it are visible to the public. However, it closely guards its image and brand, so subclassing Apple to alter and customize it will not be allowed.

4. Create a new file called `Apple`, and then create a class for Apple that uses the `public` access control:

```
public class Person {
 public let name: String
 public init(name: String) {
  self.name = name
 }
}

public class Apple {
 public private(set) var ceo: Person
 private var employees = [Person]()
 public let store = AppleStore()
 private let secretDepartment = SecretProductDepartment()

 public init() {
  ceo = Person(name: "Tim Cook")
  employees.append(ceo)
 }

 public func newEmployee(person: Person) {
  employees.append(person)
 }

 func weeklyProductMeeting() {
  var superSecretProduct = secretDepartment.
nextProduct(codeWord: "Not sure… Abracadabra?") // nil
  // Try again superSecretProduct =
  secretDepartment.nextProduct(givenCodeWord: "Titan")
  print(superSecretProduct as Any) // "Apple Glasses"
 }
}
```

5. Lastly, we have the `open` access control. Structures defined as `open` are available outside the module and can be subclassed and overridden without restriction. To explain this last control, we want to model something that exists within Apple's domain but is completely open and free from restrictions. So, for this, we can use the Swift language itself!

Swift has been open sourced by Apple, so while they maintain the project, the source code is fully available for others to take, modify, and improve.

Create a new file called `SwiftLanguage`, and then create a class for the Swift language that uses the open access control:

```
open class SwiftLanguage {
  open func versionNumber() -> Float {
    return 5.1
  }

  open func supportedPlatforms() -> [String] {
    return ["iOS", "macOS", "tvOS", "watchOS", "Linux"]
  }
}
```

We now have a module that uses Swift's access controls to provide interfaces that match our model and the appropriate visibility.

How it works...

Let's examine our `SecretProductDepartment` class to see how its visibility matches our model:

```
class SecretProductDepartment {
  private var secretCodeWord = "Titan"
  private var secretProducts = ["Apple Glasses", "Apple Car", "Apple
Brain Implant"]

  func nextProduct(codeWord: String) -> String? {
    let codeCorrect = codeWord == secretCodeWord
    return codeCorrect ? secretProducts.first : nil
  }
}
```

The `SecretProductDepartment` class is declared without an access control keyword, and when no access control is specified, the default control of `internal` is applied. Since we want the secret product department to be visible within Apple, but not from outside Apple, this is the correct access control to use.

The two properties of the `secretCodeWord` and `secretProducts` classes are marked as `private`, thus hiding their values and existence from anything outside the `SecretProductDepartment` class. To see this restriction in action, add the following to the same file, but outside the class:

```
let insecureCodeWord = SecretProductDepartment().secretCodeWord
```

When you try to build the module, you are told that `secretCodeWord` can't be accessed due to the `private` protection level.

While these properties are not directly accessible, we can provide an interface that allows information to be provided in a controlled way. This is what the `nextProduct` method provides:

```
func nextProduct(codeWord: String) -> String? {
  let codeCorrect = codeWord == secretCodeWord
  return codeCorrect ? secretProducts.first : nil
}
```

If the correct codeword is passed, it will provide the name of the next product from the secret department, but the details of all other products, and the codeword itself, will be hidden. Since this method doesn't have a specified access control, it is set to the default of `internal`.

> **Note**
>
> It's not possible for contents within a structure to have a more permissive access control than the structure itself. For instance, we can't define the `nextProduct` method as being public because this is more permissive than the class it is defined in, which is only internal.
>
> Thinking about it, this is the obvious outcome, as you cannot create an instance of an internal class outside of the defining module, so how can you possibly call a method on a class instance that you can't even create?

Now, let's look at the `AppleStore.swift` file we created. The purpose here is to provide people outside of Apple with the ability to purchase an *iPhone* through the Apple Store, restricting the creation of iPhones to just the factories where they are built, and then restricting access to those factories to just the Apple Store:

```
public enum DeviceModel {
  case iPhone13
  case iPhone13Mini
  case iPhone13Pro
  case iPhone13ProMax
}

public class AppleiPhone {
  public let model: DeviceModel
  fileprivate init(model: DeviceModel) {
    self.model = model
  }
}

public class AppleStore {
  private var factory = Factory()
  public func selliPhone(ofModel model: DeviceModel) -> AppleiPhone {
```

```
  return factory.makeiPhone(ofModel: model)
  }
}
```

Since we want to be able to sell iPhones outside of the `AppleInc` module, the `DeviceModel` enum and the `AppleiPhone` and `AppleStore` classes are all declared as `public`. This has the benefit of making them available outside the module but preventing them from being subclassed or modified. Given how Apple protects the look and feel of their phones and stores, this seems correct for this model.

The Apple Store needs to get their iPhones from somewhere – that is, from the factory:

```
fileprivate class Factory {
  func makeiPhone(ofModel model: DeviceModel) -> AppleiPhone {
    return AppleiPhone(model: model)
  }
}
```

By making the `Factory` class `fileprivate`, it is only visible within this file, which is perfect because we only want the Apple Store to be able to use the factory to create iPhones.

We have also restricted the iPhone's initialization method so that it can only be accessed from structures in this file:

```
fileprivate init(model: DeviceModel)
```

The resulting `AppleiPhone` is public, but only structures within this file can create `AppleiPhone` class objects in the first place. In this case, this is done by the factory.

Now, let's look at the `Apple.swift` file:

```
public class Person {
  public let name: String
  public init(name: String) {
    self.name = name
  }
}

public class Apple {
  public private(set) var ceo: Person
  private var employees = [Person]()
  public let store = AppleStore()
  private let secretDepartment = SecretProductDepartment()
  public init() {
    ceo = Person(name: "Tim Cook")
```

```
    employees.append(ceo)
  }

  public func newEmployee(person: Person) {
    employees.append(person)
  }

  func weeklyProductMeeting() {
    var superSecretProduct = secretDepartment.nextProduct(givenCodeWord:
"Not sure... Abracadabra?") // nil
    // Try again superSecretProduct =
    secretDepartment.nextProduct(givenCodeWord: "Titan")
    print(superSecretProduct) // "Apple Glasses"
  }
}
```

The preceding code made both the `Person` and `Apple` classes public, along with the `newEmployee` method. This allows new employees to join the company. The CEO, however, is defined as both `public` and `private`:

```
public private(set) var ceo: Person
```

We can define a separate, more restrictive, access control to set a property than the one that was set to get it. This has the effect of making it a read-only property from outside the defining structure. This provides the access we require, since we want the CEO to be visible outside of the `AppleInc` module, but we want to only be able to change the CEO from within Apple.

The final access control is `open`. We applied this to the `SwiftLanguage` class:

```
open class SwiftLanguage {
  open func versionNumber() -> Float {
    return 5.0
  }

  open func supportedPlatforms() -> [String] {
    return ["iOS", "macOSX", "tvOS", "watchOS", "Linux"]
  }
}
```

By declaring the class and methods as `open`, we allow them to be subclassed, overridden, and modified by anyone, including those outside the `AppleInc` module. With the Swift language being fully open source, this matches what we are trying to achieve.

There's more...

With our module fully defined, let's see how things look from outside the module. We need to build the module to make it available to the playground. Select the playground; it should contain a statement that imports the `AppleInc` module:

```
import AppleInc
```

First, let's look at the most accessible class that we created – that is, `SwiftLanguage`. Let's subclass the `SwiftLanguage` class and override its behavior:

```
class WinSwift: SwiftLanguage {
  override func versionNumber() -> Float {
   return 5.9
  }

  override func supportedPlatforms() -> [String] {
   var supported = super.supportedPlatforms()
   supported.append("Windows")
   return supported
  }
}
```

Since `SwiftLanguage` is open, we can subclass it to add more supported platforms and increase its version number.

Now, let's create an instance of the `Apple` class and see how we can interact with it:

```
let apple = Apple()
let keith = Person(name: "Keith Moon")
apple.newEmployee(person: keith)
print("Current CEO: \(apple.ceo.name)")
let craig = Person(name: "Craig Federighi")
apple.ceo = craig // Doesn't compile
```

We can create `Person` and provide it to Apple as a new employee, since the `Person` class and the newEmployee method are declared as public. We can retrieve information about the CEO, but we aren't able to set a new CEO, as we defined the property as `private(set)`.

Another one of the public interfaces provided by the module, `selliPhone`, allows us to buy an iPhone from the Apple Store:

```
// Buy new iPhone
let boughtiPhone = apple.store.selliPhone(ofModel: .iPhone13ProMax)
// This works
```

```
// Try and create your own iPhone
let buildAniPhone = AppleiPhone(model: .iPhone6S)
// Doesn't compile
```

We can retrieve a new iPhone from the Apple Store because we declared the `selliPhone` method as `public`. However, we can't create a new iPhone directly, since the iPhone's `init` method is declared as `fileprivate`.

See also

Further information about access control can be found in Apple's documentation on the Swift language at `http://swiftbook.link/docs/access-control`.

3

Data Wrangling with Swift

Programming is all about making decisions. The purpose of most code involves taking information, inspecting it, making decisions, and producing an output. So far, we have seen a lot of ways to represent information, but in this chapter, we will explore how to make decisions based on that information, using a number of Swift's control flow statements. We will find out how they differ and the situations where each is appropriate.

Once we've learned how Swift's control flow works, we will have opened up a world of possibilities and paths for any information we wish to work with in Swift!

In this chapter, we will cover the following recipes:

- Making decisions with if/else
- Handling all cases with switch
- Looping with for loops
- Looping with while loops
- Handling errors with try, throw, do, and catch
- Checking upfront with guard
- Doing it later with defer
- Bailing out with fatalError and precondition

Technical requirements

All the code for this chapter can be found in this book's GitHub repository at `https://github.com/PacktPublishing/Swift-Cookbook-Third-Edition/tree/main/Chapter%203`.

Making decisions with if/else

The if/else statement is a cornerstone of almost every programming language. It enables code to be executed conditionally, based on the outcome of a Boolean statement. In this recipe, we will see how if/else can be used, including some ways that are unique to Swift.

Getting ready

If you have ever played pool, you'll know that the aim of the game (when playing standard eight-ball pool) is to pot all the balls of one type and then to pot the black ball. When using American pool balls, they are numbered 1–15 and have a different pattern, depending on their type. Balls 1–7 have a solid color, balls 9–15 are white with a colored stripe around them, and ball 8 is black:

Figure 3.1 – American pool balls

In this recipe, we will write a function that will take the number on a pool ball and return the type of ball it is.

Let's create a new playground for this recipe.

How to do it...

Let's use an if/else control flow statement to write a function to return the right pool ball type:

1. Create an enum to describe the possible ball types:

    ```
    enum PoolBallType {
    case solid
    case stripe
    case black
    }
    ```

2. Create the method that will take Int and return PoolBallType:

```
func poolBallType(forNumber number: Int) -> PoolBallType {
    if number < 8 {
        return .solid
    } else if number > 8 {
        return .stripe
    } else {
        return .black
    }
}
```

3. Use this function, and check that we get the expected results:

```
let two = poolBallType(forNumber: 2) // .solid
let eight = poolBallType(forNumber: 8) // .black
let twelve = poolBallType(forNumber: 12) // .stripe
```

How it works...

Within the function, we define three code paths – `if`, `else if`, and `else`:

```
if <#a boolean expression#> {
    <#executed if boolean expression above is true#>
} else if <#other boolean expression#> {
    <#executed if other boolean expression above is true#>
} else {
    <#executed if neither boolean expressions are true#>
}
```

First, we want to determine whether the ball is solid. Since we know that the balls numbered 1–7 are solid, we can test whether the ball number is less than 8, with number <8. If this is `true`, we return the `.solid` case of our enum. If it is `false`, the `else if` Boolean expression is evaluated.

As balls 9–15 are striped, we can test whether the ball number is more than 8, with number > 8. If this is `true`, we return the `.stripe` case of our enum.

Lastly, if both the preceding Boolean expressions are `false`, we return the `.black` case of our enum, since that can only happen if the number is exactly 8.

The `else if` and `else` blocks are optional, and you can declare multiple `else if` to cover additional conditions. Let's expand our preceding example with an extra `else if` to better decide the pool ball type.

As we stated previously, pool balls are numbered between 1 and 15, but we don't take into account those upper and lower bounds in our implementation. So, if we were to provide the function with ball number 0, it would return `.solid`, and if we were to provide ball number 16, it would return `.stripe`, which doesn't accurately reflect our intention:

```
let zero = poolBallType(forNumber: 0) // .solid
let sixteen = poolBallType(forNumber: 16) // .stripe
```

Let's modify our function to only return a pool ball type if the number is between 1 and 15, returning `nil` otherwise:

```
func poolBallType(forNumber number: Int) -> PoolBallType? {
    if number > 0 && number < 8 {
        return .solid
    } else if number > 8 && number < 16 {
        return .stripe
    } else if number == 8 {
        return .black
    } else {
        return nil
    }
}
```

Now, we have four code branches in our `if` statement, and we can use the AND operator, `&&`, to combine Boolean statements (the OR operator, `||`, is also available).

We can now call our function for both numbers within the expected 1–15 range and outside it:

```
let two = poolBallType(forNumber: 2) // .solid
let eight = poolBallType(forNumber: 8) // .black
let twelve = poolBallType(forNumber: 12) // .stripe
let zero = poolBallType(forNumber: 0) // nil
let sixteen = poolBallType(forNumber: 16) // nil
```

Our improved function will produce `nil` for numbers outside of the expected range.

There's more...

There are some other ways we can use `if`/`else` statements.

Understanding conditional unwrapping

The function we created earlier returns an optional value, so if we want to do anything useful with the resulting value, we need to unwrap the optional. So far, the only way we have seen how to do this is by force-unwrapping, which will cause a crash if the value is `nil`.

Instead, we can use an `if` statement to *conditionally unwrap* the optional, turning it into a more useful, non-optional value.

Let's create a function that will print information about a pool ball of a given number. If the provided number is valid for a pool ball, it will print the ball's number and type; otherwise, it will print a message explaining that it is not a valid number.

Since we will want to print the value of the `PoolBallType` enum, let's make it `String`-backed, which will make printing its value easier:

```
enum PoolBallType: String {
      case solid
      case stripe
      case black
}
```

Now, let's write the function to print the pool ball details:

```
func printBallDetails(ofNumber number: Int) {
      let possibleBallType = poolBallType(forNumber: number)
      if let ballType = possibleBallType {
            print("\(number) - \(ballType.rawValue)")
      } else {
            print("\(number) is not a valid pool ball number")
      }
}
```

The first thing we do in our `printBallDetails` function is get the ball type for the given number:

```
let possibleBallType = poolBallType(forNumber: number)
```

In our improved version of this function, this returns an optional version of the `PoolBallType` enum. We want to include the `rawValue` of the returned enum as part of printing the ball details. Since the returned value is optional, we need to unwrap it first:

```
if let ballType = possibleBallType {
      print("\(number) - \(ballType.rawValue)")
}
```

In the preceding `if` statement, instead of defining a Boolean expression, we assign our optional value to a constant; the `if` statement uses this to *conditionally unwrap* the optional value. The optional value is checked to see whether it is `nil`; if it is not `nil`, then the value is unwrapped and assigned to the constant as a non-optional value. That constant becomes available within the scope of the curly brackets following the `if` statement. We use that `ballType` non-optional value to obtain the raw value for the `print` statement.

Since the `if` branch of the `if/else` statement is followed when the optional value has a value, then the `else` branch is followed when the optional value is `nil`.

As this means that the given number is not valid for a pool ball, we print a relevant message:

```
else {
        print("\(number) is not a valid pool ball number")
}
```

We can now call our new function with the same values as before to print out the pool ball type:

```
printBallDetails(ofNumber: 2) // 2 - solid
printBallDetails(ofNumber: 8) // 8 - black
printBallDetails(ofNumber: 12) // 12 - stripe
printBallDetails(ofNumber: 0) // 0 is not a valid pool ball number
printBallDetails(ofNumber: 16) // 16 is not a valid pool ball number
```

We've used conditional unwrapping to print the pool ball type, if valid, or explain if it's not valid.

> **Note**
>
> Note that we unwrap our optional into a newly named constant. Even though it's only available within the scope of the if statement, we now have to track our value under a new name, which can be confusing. However, as of Swift 5.7, we can now use a new let shorthand to shadow our optional value.
>
> Therefore, we can write the preceding example like this:
>
> ```
> if let possibleBallType = poolBallType(forNumber: ballNumber) {
>
>     print("\(ballNumber) - \(possibleBallType.rawValue)")
>
> }
> ```

Chaining optional unwrapping

The ability of `if` statements to conditionally unwrap optionals can be chained together to produce some useful and concise code. The following example is a bit contrived, but it illustrates how we can use a single `if` statement to unwrap a chain of optional values.

When you play a game of pool, called a *frame*, the type of the first ball you pot becomes the type you need to pot for the rest of the frame, and your opponent has to pot the opposite type.

Let's define a frame of pool and say that we want to track what type of ball each player will be potting:

```
class PoolFrame {
        var player1BallType: PoolBallType?
        var player2BallType: PoolBallType?
}
```

We will also create a `PoolTable` object that has an optional `currentFrame` property, which will contain information about the current frame if one is in progress:

```
class PoolTable {
        var currentFrame: PoolFrame?
}
```

We now have a pool table that has an optional frame and a frame that has an optional ball   type for each player.

Now, let's write a function that prints the ball type for player 1 in the current frame. It is possible that the current frame is `nil` because there is no frame currently being played, or that player 1's ball type is `nil` because a ball hasn't yet been potted. Therefore, we need to account for either of those values being `nil`:

```
func printBallTypeOfPlayer1(forTable table: PoolTable) {
        if let frame = table.currentFrame, let ballType = frame.
player1BallType {
                print(ballType.rawValue)
        } else {
                print("Player 1 has no ball type or there is no current
frame")
        }
}
```

Our function is given `PoolTable`, and to print player 1's ball type, we first need to check and unwrap the `currentFrame` property, and then we need to check and unwrap the current frame's `player1BallType` property.

We can do this by nesting our `if` statements:

```
func printBallTypeOfPlayer1(forTable table: PoolTable) {
        if let frame = table.currentFrame {
                if let ballType = frame.player1BallType {
                        print(ballType.rawValue)
                } //... handle else
        } //... handle else
}
```

Alternatively, we can handle this chained unwrapping in one `if` statement by performing the unwrapping statement sequentially, separated by commas, and ensuring that each statement can access the unwrapped values from the previous statements:

```
func printBallTypeOfPlayer1(forTable table: PoolTable) {
        if let frame = table.currentFrame, let ballType = frame.
player1BallType {
                print("\(ballType)")
        } //... handle else
}
```

The first statement unwraps the `currentFrame` property, and the second statement uses that unwrapped frame to unwrap player 1's ball type.

Let's use the function we've just created:

1. First, we'll create a table, without a current frame, and print player 1's ball type, which won't be available:

    ```
    //
    // Table with no frame in play
    //
    let table = PoolTable()
    table.currentFrame = nil
    printBallTypeOfPlayer1(forTable: table)
    // Player 1 has no ball type or there is no current frame
    ```

2. Then, we can create a current frame, but as player 1's ball type is still nil, the function prints the same output:

    ```
    //
    // Table with frame in play, but no balls potted
    //
    let frame = PoolFrame()
    frame.player1BallType = nil
    frame.player2BallType = nil
    table.currentFrame = frame
    printBallTypeOfPlayer1(forTable: table)
    // Player 1 has no ball type or there is no current frame
    ```

3. If we set player 1's ball type, our function now prints the type:

    ```
    //
    // Table with frame in play, and a ball potted
    //
    frame.player1BallType = .solid
    ```

```
frame.player2BallType = .stripe
printBallTypeOfPlayer1(forTable: table)
// solid
```

We've created a method that can chain conditional unwrappings, only printing a value when all the values in the chain are non-`nil`.

Using enums with associated values

As we saw in the *Enumerating values with enums* recipe of *Chapter 1*, enums can have associated values, and we can use an `if` statement to both check an enum's case and extract the associated value in one expression.

Let's create an enum to represent the result of the pool game, with each case having an associated message:

```
enum FrameResult {
        case win(congratulations: String)
        case lose(commiserations: String)
}
```

Now, we'll create a function that takes `Result` and prints either the congratulatory message or the commiseration message:

```
func printMessage(forResult result: FrameResult) {
        if case Result.win(congratulations: let winMessage) = result {
                print("You won! \(winMessage)")
        } else if case Result.lose(commiserations: let loseMessage) =
result {
                print("You lost :( \(loseMessage)")
        }
}
```

Calling this function will print the result, followed by the relevant message:

```
let result = Result.win(congratulations: "You're simply the best!")
printMessage(forResult: result) // You won! You're simply the best!
```

The `if` case block will be executed if the value on the right-hand side of = matches the case on the left-hand side. In addition, you can specify a local constant for the associated value (`winMessage` in the following example), which is then available within the subsequent block:

```
if case Result.win(congratulationsMessage: let winMessage) = result {
        print("You won! \(winMessage)")
}
```

We've used the `if` case statement to both check the case of an `enum` value and access its associated value in one go.

See also

Further information about `if`/`else` can be found in Apple's documentation on the Swift language at `http://swiftbook.link/docs/statements`.

Handling all cases with switch

`switch` statements allow you to control the flow of execution by testing one specific value in multiple ways. In Objective-C and other languages, `switch` statements can only be used on values that can be represented by an integer, and they are most commonly used to make decisions based on enumeration cases.

As we have seen, **enumerations** have become a lot more powerful in Swift, as they can be based on more than just integers, and so too can `switch` statements.

`switch` statements in Swift can be used on any type and have advanced pattern-matching functionality.

In this recipe, we will explore both simple and advanced usage of `switch` control flow statements to control logic.

Getting ready

If you are old enough to remember the early days of the home computer, you may also remember text-based adventures. These were simple games that usually described a scene and then let you move around, by typing a command to move north, south, east, or west. You would find and pick up items, and you could often combine them to solve puzzles.

We can use `switch` statements to control the logic of a simple text adventure.

Let's create a new playground for this recipe.

How to do it...

Let's create parts of a text-based adventure and use `switch` statements to make the decisions:

1. Define an enum to represent the directions we can travel in:

    ```
    enum CompassPoint {
        case north
        case south
        case east
        case west
    }
    ```

2. Create a function that describes what the player of the text adventure will see when they look in that direction:

```
func lookTowards(_ direction: CompassPoint) {
        switch direction {
        case .north:
                print("To the north lies a winding road")
        case .south:
                print("To the south is the Prancing Pony tavern")
        case .east:
                print("To the east is a blacksmith")
        case .west:
                print("To the west is the town square")
        }
}
lookTowards(.south) // To the south is the Prancing Pony tavern
```

3. In our text adventure, users can pick up items and attempt to combine them, to produce new items and solve problems. Define our available items as an enum:

```
enum Item {
        case key
        case lockedDoor
        case openDoor
        case bluntKnife
        case sharpeningStone
        case sharpKnife
}
```

4. Write a function that takes two items and tries to combine them into a new item. If the items cannot be combined, it will return nil:

```
func combine(_ firstItem: Item, with secondItem: Item) -> Item?
{
        switch (firstItem, secondItem) {
        case (.key, .lockedDoor):
                print("You have unlocked the door!")
                return .openDoor
        case (.bluntKnife, .sharpeningStone):
                print("Your knife is now sharp")
                return .sharpKnife
        default:
                print("\(firstItem) and \(secondItem) cannot be
combined")
                return nil
        }
```

```
    }
    let door = combine(.key, with: .lockedDoor) // openDoor
    let oilAndWater = combine(.bluntKnife, with: .lockedDoor) // nil
```

5. In our text adventure, the player will meet different characters and can interact with them. Define the characters that the player can meet:

```
    enum Character: String {
          case wizard
          case bartender
          case dragon
    }
```

6. Write a function that allows the player to say something, and optionally provide a character to whom it will be said. The interaction that will occur will depend on what is said and the character it is said to:

```
    func say(_ textToSay: String, to character: Character? = nil) {
          switch (textToSay, character) {
          case ("abracadabra", .wizard?):
                print("The wizard says, \"Hey, that's my
    line!\"")
          case ("Pour me a drink", .bartender?):
                print("The bartender pours you a drink")
          case ("Can I have some of your gold?", .dragon?):
                print("The dragon burns you to death with his
    fiery breath")
          case (let textSaid, nil):
                print("You say \"\(textSaid)\", to no-one.")
          case (_, let anyCharacter?):
                print("The \(anyCharacter) looks at you,
    blankly")
          }
    }

    say("Is anybody there?")
    // You say "Is anybody there?", to no-one.
    say("Pour me a drink", to: .bartender)
    // The bartender pours you a drink
    say("Can I open a tab?", to: .bartender)
    // The bartender looks at you, blankly
```

How it works...

Within the `lookTowards` function, we want to print a different message for each possible `CompassPoint` case; to do this, we use a `switch` statement:

```swift
func lookTowards(_ direction: CompassPoint) {
    switch direction {
    case .north:
        print("To the north lies a winding road")
    case .south:
        print("To the south is the Prancing Pony tavern")
    case .east:
        print("To the east is a blacksmith")
    case .west:
        print("To the west is the town square")
    }
}
```

At the top of the `switch` statement, we define the value that we want to switch on; then, we define what we want to be done when that value matches each of the defined cases, using the `case` keyword and then the matching pattern:

```swift
switch <#value#> {
case <#pattern#>:
    <#code#>
case <#pattern#>:
    <#code#>
//...
}
```

Each `case` statement is evaluated in turn, and if the pattern matches the value, the subsequent code is executed.

> **Note**
>
> If you are familiar with switch statements from Objective-C, you may remember that you needed to add break; at the end of each case statement to stop the execution from falling through to the next case statement. This is not needed in Swift; the break in execution is implied by the beginning of the next case statement. The only time this isn't the case is when your case statement is intentionally empty; in these cases, you need to add break to tell the compiler that it is intentionally blank for this case. If you do want execution to fall through to the next case statement, you can add fallthrough at the end of the case statement.

In our `combine` function, we have two values that we want to switch based on their values. We can provide multiple values to the `switch` statement in the form of a tuple:

```
func combine(_ firstItem: Item, with secondItem: Item) -> Item? {
    switch (firstItem, secondItem) {
        //....
    }
}
```

For each `case` statement, we define the valid value for each part of the tuple:

```
case (.key, .lockedDoor):
    print("You have unlocked the door!")
    return .openDoor
```

A `switch` statement in Swift requires that every possible case is covered; however, you can cover all the remaining possibilities in one go using the `default` case:

```
switch (firstItem, secondItem) {
//...
default:
    print("\(firstItem) and \(secondItem) cannot be combined")
    return nil
}
```

For our preceding `combine` function, you will notice that the player will only be able to combine the items if they provide them in the right order:

```
let door1 = combine(.key, with: .lockedDoor) // openDoor
let door2 = combine(.lockedDoor, with: .key) // nil
```

This is not the desired behavior, as there is no way for the player to know the correct order. To solve this, we can add multiple patterns to each `case` statement. So, when the player provides the `key` and `lockedDoor` items, we can handle the `key` and `lockedDoor` order and the `lockedDoor` and `key` order with the same `case` statement, using the following format:

```
switch <#value#> {
case <#pattern#>, <#pattern#>:
    <#code#>
default:
    <#code#>
}
```

So, we can add the opposite item order as another pattern to each case:

```
func combine(_ firstItem: Item, with secondItem: Item) -> Item? {
    switch (firstItem, secondItem) {
    case (.key, .lockedDoor), (.lockedDoor, .key):
        print("You have unlocked the door!")
        return .openDoor
    case (.bluntKnife, .sharpeningStone),
        (.sharpeningStone, .bluntKnife):
        print("Your knife is now sharp")
        return .sharpKnife
    default:
        print("\(firstItem) and \(secondItem) cannot be
combined")
        return nil
    }
}
```

Now, the items can be combined in any order:

```
let door1 = combine(.key, with: .lockedDoor) // openDoor
let door2 = combine(.lockedDoor, with: .key) // openDoor
```

For our say method, we again have two values that we want to switch on – the text that the player says, and the character to whom it is said. Since the character value is optional, we will need to unwrap the value to compare it with non-optional values:

```
func say(_ textToSay: String, to character: Character? = nil) {
    switch (textToSay, character) {
    case ("abracadabra", .wizard?):
        print("The wizard says, "Hey, that's my line!"")
        //...
    }
}
```

In a switch statement, when the value is optional, you can compare it to a non-optional value by adding ? to wrap it as an optional, making the comparison valid. In the preceding instance, we compare the optional character value to .wizard?.

Where we have two values for a certain set of options, we may only care about one of the values, and the other value could be anything and the case would still be valid. In our example, once all the specific `textToSay` and character pairings have been handled, and the case where there is no character is handled, we want to unwrap and retrieve the character, but we don't care about the `textToSay` value, so we can use _ to indicate that any value is acceptable:

```
func say(_ textToSay: String, to character: Character? = nil) {
    switch (textToSay, character) {
    //...
    case (_, let anyCharacter?):
            print("The \(anyCharacter) looks at you, blankly)")
    }
}
```

To retrieve the value of the character entered as part of this `case` statement rather than declaring a value to be matched, we define a constant that will receive the value, and since the value we are switching on is optional, we also add ?, which will unwrap the value if it's not `nil`, and assign it to the constant.

See also

Further information about `switch` can be found in Apple's documentation on the Swift language at `http://swiftbook.link/docs/switch`.

Looping with for loops

`for` loops allow you to execute code for each element in a collection or range. In this recipe, we will explore how to use `for` loops to perform actions on every element in a collection.

Getting ready

In this recipe, we won't be using any components from the previous recipes, so you can create a new playground for this recipe.

How to do it...

Let's create some collections and then use `for` loops to act on each element in the collection:

1. Create an array of elements so that we can do something with every item in the array:

    ```
    let ledZeppelin = ["Robert", "Jimmy", "John", "John Paul"]
    ```

2. Create a loop to go through our theBeatles array, and print each string element that the for loop provides:

```
for musician in ledZeppelin {
        print(musician)
}
```

3. Create a for loop that executes some code a set number of times, instead of looping through an array. We can do this by providing a range instead of a collection:

```
// 5 times table
for value in 1...12 {
        print("5 x \(value) = \(value*5)")
}
```

4. Create a for loop to print the keys and values of a dictionary. Dictionaries contain pairings between a key and a value, so when looping through a dictionary, we will be provided with both the key and the value in the form of a tuple:

```
let zeppelinByInstrument = ["vocals": "Robert",
                            "lead guitar": "Jimmy",
                            "drums": "John",
                            "bass guitar": "John Paul"]
for (key, value) in zeppelinByInstrument {
        print("\(value) plays \(key)")
}
```

How it works...

Let's look at how we looped through our `ledZeppelin` array:

```
for musician in ledZeppelin {
        print(musician)
}
```

We specify the `for` keyword, and then we provide a name for the local variable that will be used for each element in the collection or range. Then, the `in` keyword is provided, followed by the collection or range that will be looped through:

```
for <#each element#> in <#collection or range#> {
        <#code to execute#>
}
```

For range-based loops, the value provided for each loop is the next integer in the range:

```
for value in 1...12 {
      print("5 x \(value) = \(value*5)")
}
```

A range can be a *closed range*, where the range includes the start value and the end value, like the one specified previously. Alternatively, it can be a *half-open range*, which goes up to, but doesn't include, the last value, as in the following code:

```
for value in 1..<13 {
      print("5 x \(value) = \(value*5)")
}
```

When looping through a dictionary, we need to be provided with both the key and value properties. To do this, we provide a tuple that will receive each key and value in the dictionary:

```
for (key, value) in zeppelinByInstrument {
      print("\(value) plays \(key)")
}
```

We can define the tuple and name each of the values. This name can then be used in an execution block. Let's change the tuple labels to better describe the values:

```
for (instrument, musician) in zeppelinByInstrument {
      print("\(musician) plays \(instrument)")
}
```

Giving the tuple meaningful names in the preceding example makes the code easier to read.

> **Note**
>
> An alternative way to loop through each element in a sequence is to call the forEach method directly on the sequence itself:
>
> ```
> ledZeppelin.forEach { musician in
>       print(musician)
> }
> zeppelinByInstrument.forEach { (key, value) in
>       print(«\(value) plays \(key)»)
> }
> ```

See also

Further information about for-in loops can be found in Apple's documentation on the Swift language at `http://swiftbook.link/docs/for-in`.

Looping with while loops

for loops are great when you know how many times you intend to loop, but if you want to loop until a certain condition is met, you need a while loop.

A while loop has the following syntax:

```
while <#boolean expression#> {
        <#code to execute#>
}
```

The code block will execute over and over until the Boolean expression returns false. Therefore, it's a common pattern to change some value in the code block that may cause the Boolean expression to change to false.

> **Note**
>
> If there is no chance of the Boolean expression becoming true, the code will loop forever, which can lock up your app.

In this recipe, we will look at situations where a while loop can be useful to repeat actions.

Getting ready

This recipe will involve simulating the random flip of a coin. To flip our coin, we will need to randomly pick either heads or tails, so we will need to use a random number generator from the **Foundation** framework. We will discuss Foundation further in *Chapter 5*, *Beyond the Standard Library*, but for now, let's just import the Foundation framework at the top of a new playground:

```
import Foundation
```

This will give us the ability to generate a random number to use in this recipe.

How to do it...

Let's work out how many times in a row we can flip a coin and get heads:

1. Create an enum to represent a coin flip, and use the random number generator to randomly choose heads or tails:

    ```swift
    enum CoinFlip: Int {
            case heads
            case tails
            static func flipCoin() -> CoinFlip {
                    return CoinFlip(rawValue: Int(arc4random_
    uniform(2)))!
            }
    }
    ```

2. Create a function that will return the number of heads in a row from coin flips. The function will flip the coin within a while loop and continue to loop while the coin flip results in heads:

    ```swift
    func howManyHeadsInARow() -> Int {
            var numberOfHeadsInARow = 0
            var currentCoinFlip = CoinFlip.flipCoin()
            while currentCoinFlip == .heads {
                    numberOfHeadsInARow = numberOfHeadsInARow + 1
                    currentCoinFlip = CoinFlip.flipCoin()
            }
            return numberOfHeadsInARow
    }

    let noOfHeads = howManyHeadsInARow()
    ```

How it works...

In our function, we start by keeping track of how many coin flips in a row are heads and keep a reference to the current coin flip, which will form the condition for the `while` loop:

```swift
func howManyHeadsInARow() -> Int {
        var numberOfHeadsInARow = 0
        var currentCoinFlip = CoinFlip.flipCoin()
        //...
}
```

In our `while` loop, we will continue to loop and execute the code in the following block while the current coin flip is heads:

```
while currentCoinFlip == .heads {
     numberOfHeadsInARow = numberOfHeadsInARow + 1
     currentCoinFlip = CoinFlip.flipCoin()
}
```

Within the code block, we add one to our running total, and we re-flip the coin. We flip the coin and assign it to `currentCoinFlip`, which will get rechecked on the next loop, and if it is still heads, the next loop will be executed. Since we are changing something that affects the `while` condition, such that it could eventually be `false`, we can be sure that we won't be stuck in the loop forever.

As soon as the coin flip is tails, the `while` loop condition will be `false`, and so the execution will move on and return the running total we have been keeping:

```
return numberOfHeadsInARow
```

Now, every time you call the function, the coin will be randomly flipped, and the number of heads in a row will be returned, so each time it's called, you may get a different value returned. Try it out a few times:

```
let noOfHeads = howManyHeadsInARow()
```

There's more...

We can actually simplify our `while` loop by doing the coin flip as part of the loop continuation checking:

```
func howManyHeadsInARow() -> Int {
     var numberOfHeadsInARow = 0
     while CoinFlip.flipCoin() == .heads {
          numberOfHeadsInARow = numberOfHeadsInARow + 1
     }
     return numberOfHeadsInARow
}
```

Each time through the loop, the `while` condition is evaluated, which involves re-flipping the coin and checking the outcome.

This is more concise and removes the need to track `currentCoinFlip`.

See also

Further information about `while` loops can be found in Apple's documentation on the Swift language at `http://swiftbook.link/docs/while`.

Handling errors with try, throw, do, and catch

Errors happen during programming. These errors may be due to your own code behaving in unexpected ways, or due to unexpected information or behavior from external systems. When these errors happen, it's important to handle them appropriately. Good error handling can separate a good app from a great app.

Swift provides a deliberate and flexible pattern to handle errors, allowing specific errors to be cascaded through a complex system.

In this recipe, we will discover how to define errors and throw them when necessary.

Getting ready

In this recipe, we won't be using any components from the previous recipes, so you can create a new playground for this recipe.

How to do it...

To examine error handling, we will model a process that can go wrong, and for me, that is cooking a meal:

1. First, let's define the steps involved in cooking a meal as states that the meal will transition through:

    ```
    enum MealState {
            case initial
            case buyIngredients
            case prepareIngredients
            case cook
            case plateUp
            case serve
    }
    ```

2. Create an object to represent the meal we will be cooking. This object will hold the state of the meal as it moves through the process:

    ```
    class Meal {
            var state: MealState = .initial
    }
    ```

    We want to allow the meal to transition between states, but not all state transitions should be possible. For instance, you can't move from buying ingredients to serving the meal. The meal should move sequentially from one state to the next. We can provide these restrictions by only allowing the state to be set from within the object itself, using access controls, which we explored in the previous chapter.

3. Define the `state` property as only being privately settable:

```
class Meal {
      private(set) var state: MealState = .initial
}
```

4. To allow the state to be changed from outside the object, create a function inside the class that will throw an error if the state transition isn't possible:

```
func change(to newState: MealState) throws {
      switch (state, newState) {
      case (.initial, .buyIngredients),
            (.buyIngredients, .prepareIngredients),
            (.prepareIngredients, .cook),
            (.cook, .plateUp),
            (.plateUp, .serve):
                state = newState
      default:
            throw MealError.canOnlyMoveToAppropriateState
      }
}
```

5. In keeping with Swift's protocol-orientated approach, errors in Swift are defined as a protocol, `Error`. This approach allows you to construct your own type to represent errors within your code, and you just have it conform to the `Error` protocol.

    A common approach is to define errors as enums, with the enum cases representing the different types of errors that can occur.

    Define the error thrown in the preceding `Meal` class:

```
enum MealError: Error {
      case canOnlyMoveToAppropriateState
}
```

6. Try to execute our error-throwing method within a do block, and catch any errors that may occur:

```
let dinner = Meal()
do {
      try dinner.change(to: .buyIngredients)
      try dinner.change(to: .prepareIngredients)
      try dinner.change(to: .cook)
      try dinner.change(to: .plateUp)
      try dinner.change(to: .serve)
      print("Dinner is served!")
```

```
} catch let error {
        print(error)
}
```

How it works...

The terminology used in Swift error handling (as well as other languages) is *throwing* and *catching*. A method can *throw* an error if a problem occurs during its execution, at which point nothing further in the method will be executed, and the error is passed back to where the method was called from.

In order to receive this error (perhaps to provide the details of the error to the user), you must *catch* the error at the place the method is called.

To throw an error, you have to declare that the method has the potential to throw an error. Declaring that a method throws allows the compiler to expect potential errors from the method, and ensure that you don't forget to catch these errors.

Methods can be declared as potentially throwing an error using the `throws` keyword:

```
func change(to newState: MealState) throws {
        //...
}
```

Within our change state method, we only change the state if we are moving to the next sequential state. Anything else isn't allowed and should throw an error. We can do this using the `throw` keyword, followed by a value that conforms to the `Error` protocol:

```
func change(to newState: MealState) throws {
        //...
        default:
                throw MealError.canOnlyMoveToAppropriateState
        }
}
```

When we create the `Meal` object and move through the states of preparing the meal, each change of state can throw an error. When we call a method that is marked as possibly throwing an error, we have to do it a certain way. We define a `do` block, within which we can call methods that can throw, and we then define a `catch` block that will be executed if any of these methods do throw an error. Each call to a throwing method must be prefixed with the `try` keyword:

```
let dinner = Meal()
do {
        try dinner.change(to: .buyIngredients)
        try dinner.change(to: .prepareIngredients)
        try dinner.change(to: .cook)
```

```
        try dinner.change(to: .plateUp)
        try dinner.change(to: .serve)
        print("Dinner is served!")
} catch let error {
        print(error)
}
```

If any of these methods does throw an error, execution will immediately move to the `catch` block. Therefore, by placing code after the `try` methods are called, we guarantee that it will only be executed if the methods do not throw an error. By printing `Dinner is served!` after all the state transitions are called, we know this will only print if we have successfully moved through all the states. Try changing the order of these state change calls, and you'll see that the error is printed, and `Dinner is served!` is not.

In our `catch` block, after the `catch` keyword, we can define the local constant that we want the caught error to be assigned to. However, if we don't specify a local constant here, Swift will implicitly create one for us called `error`, so we can actually omit the constant declaration in the `catch` block and still print the value of the error:

```
do {
        //...
} catch {
        print(error)
}
```

Swift has defined the error for us, so we can still print the value.

There's more...

We have seen how we can throw and catch errors, but we mentioned in the introduction that we can cascade errors through a system, so let's look at how we can do this.

In our meal preparation example, we allow the meal state to be changed externally through a change method that can throw an error. Instead, let's change it to a private method, so we can only call it from within the class:

```
class Meal {
        private(set) var state: MealState = .initial
        private func change(to newState: MealState) throws {
                switch (state, newState) {
                case (.initial, .buyIngredients),
                        (.buyIngredients, .prepareIngredients),
                        (.prepareIngredients, .cook),
                        (.cook, .plateUp),
                        (.plateUp, .serve):
```

```
                        state = newState
            default:
                    throw MealError.canOnlyMoveToAppropriateState
            }
        }
    }
```

Now, let's create some specific methods to move to each state:

```
class Meal {
        //...
        func buyIngredients() throws {
                try change(to: .buyIngredients)
        }
        func prepareIngredients() throws {
                try change(to: .prepareIngredients)
        }
        func cook() throws {
                try change(to: .cook)
        }
        func plateUp() throws {
                try change(to: .plateUp)
        }
        func serve() throws {
                try change(to: .serve)
        }
}
```

Note that when we call the change method from within each of the new methods, we don't need to use do and catch blocks to catch the error; this is because we have defined each of the new methods as potentially throwing an error. So, if the call to the change method throws an error, this error will be passed to the caller of our new method as though it were throwing an error.

This mechanism allows errors that may occur several levels deep in your code to surface and be handled appropriately.

We now need to amend our meal preparation code to use these new methods:

```
let dinner = Meal()
do {
        try dinner.buyIngredients()
        try dinner.prepareIngredients()
        try dinner.cook()
        try dinner.plateUp()
```

```
        try dinner.serve()
        print("Dinner is served!")
} catch let error {
        print(error)
}
```

Let's add the ability to actually affect our meal. We'll add a method to add `salt` to the meal and a property to allow us to track how much salt is added. Add these to the end of the `Meal` class:

```
class Meal {
        //...
        private(set) var saltAdded = 0 func addSalt() throws {
                if saltAdded >= 5 {
                        throw MealError.tooMuchSalt
                } else if case .initial = state, case .buyIngredients =
state {
                        throw MealError.wrongStateToAddSalt
                } else {
                        saltAdded = saltAdded + 1
                }
        }
}
```

There are two ways in which adding salt can throw an error, either because we are in the wrong state to add `salt` (we can't add salt until after we have bought the ingredients), or because we have added too much salt. Let's add these two new errors to our `MealError` enum:

```
enum MealError: Error {
        case canOnlyMoveToAppropriateState
        case tooMuchSalt
        case wrongStateToAddSalt
}
```

We now have three possible errors that can occur during the preparation of a meal, and we may want to handle those errors differently. We can use multiple `catch` blocks to filter just specific errors that we want to catch, allowing us to handle each error separately:

```
let dinner = Meal()
do {
        try dinner.buyIngredients()
        try dinner.prepareIngredients()
        try dinner.cook()
        try dinner.plateUp()
        try dinner.serve()
        print("Dinner is served!")
```

```
    } catch MealError.canOnlyMoveToAppropriateState {
        print("It's not possible to move to this state")
    } catch MealError.tooMuchSalt {
        print("Too much salt!")
    } catch MealError.wrongStateToAddSalt {
        print("Can't add salt at this stage")
    } catch {
        print("Some other error: \(error)")
    }
}
```

> **Note**
>
> It is important to ensure that all possible errors are handled by the catch blocks, as an unhandled error will result in a crash. It is, therefore, safest to add an unfiltered catch block at the end to catch any errors not caught by the previous blocks.

Since functions can throw an error, and closures are a type of function that can be passed as a parameter, we can have a function that takes a throwing closure where it can also throw an error. It may be that the only errors our function will throw are errors produced by the throwing closure that was passed as a parameter.

When that is `true`, a function can be defined as re-throwing, using the `rethrows` keyword. This situation is quite confusing, so let's look at an example:

```
func makeMeal(using preparation: (Meal) throws -> ()) rethrows -> Meal
{
    let newMeal = Meal()
    try preparation(newMeal)
    return newMeal
}
```

This `makeMeal` function takes a closure as a parameter; that closure takes a `Meal` object as a parameter and doesn't return anything, but it may throw an error.

The purpose of this function is to handle the creation of the `meal` object for you, just leaving you to do any meal preparation within the block; it then returns the meal that was created and prepared. Let's see it in use:

```
do {
    let dinner = try makeMeal { meal in
        try meal.buyIngredients()
        try meal.prepareIngredients()
        try meal.cook()
        try meal.addSalt()
        try meal.plateUp()
```

```
            try meal.serve()
        }
        if dinner.state == .serve {
            print("Dinner is served!")
        }
} catch MealError.canOnlyMoveToAppropriateState {
        print("It's not possible to move to this state")
} catch MealError.tooMuchSalt {
        print("Too much salt!")
} catch MealError.wrongStateToAddSalt {
        print("Can't add salt at this stage")
}
```

The makeMeal function only throws errors thrown by the closure parameter, so it can be declared as re-throwing. Declaring a function of this type with the rethrows keyword isn't required; it can be declared with throws instead. However, the compiler can make additional optimizations for a re-throwing function.

See also

Further information about error handling can be found in Apple's documentation on the Swift language at http://swiftbook.link/docs/error-handling.

Checking upfront with guard

We have seen in previous recipes how we can use if statements to check Boolean expressions and unwrap optional values. It's a common use case to want to do some checks and conditional unwrapping at the beginning of a block of code, and then only execute the subsequent code if everything is as expected. This usually results in wrapping the whole block of code in an if statement:

```
if <#boolean check and unwrapping#> {
        <#a block of code#>
        <#that could be quite long#>
}
```

Swift has a better solution expressly for this purpose – the guard statement.

In this recipe, we will learn how to use the guard statement to return early from a method.

Getting ready

Let's imagine that we have some data that came from an external source, and we want to turn it into model objects that our code can understand, with the intention of displaying it to the user. We can use guard statements to ensure the data is correctly formatted, bailing early if it isn't correct.

Let's create a new playground for this recipe.

How to do it...

We will take some information about the planets of the solar system, which could have come from an external source, and turn it into a model we can understand:

1. Create the planet data in the form of an array of dictionaries:

```
// From https://en.wikipedia.org/wiki/Solar_System

let inputData: [[String: Any]] = [
["name": "Mercury", "positionFromSun": 1,
"fractionOfEarthMass": 0.055,
"distanceFromSunInAUs": 0.4, "hasRings": false],

["name": "Venus", "positionFromSun": 2,
"fractionOfEarthMass": 0.815,
"distanceFromSunInAUs": 0.7, "hasRings": false],

["name": "Earth", "positionFromSun": 3,
"fractionOfEarthMass": 1.0,
"distanceFromSunInAUs": 1.0, "hasRings": false],

["name": "Mars", "positionFromSun": 4,
"fractionOfEarthMass": 0.107,
"distanceFromSunInAUs": 1.5, "hasRings": false],

["name": "Jupiter", "positionFromSun": 5,
"fractionOfEarthMass": 318.0,
"distanceFromSunInAUs": 5.2, "hasRings": false],

["name": "Saturn", "positionFromSun": 6,
"fractionOfEarthMass": 95.0,
"distanceFromSunInAUs": 9.5, "hasRings": true],

["name": "Uranus", "positionFromSun": 7,
"fractionOfEarthMass": 14.0,
"distanceFromSunInAUs": 19.2, "hasRings": false],
```

```
["name": "Neptune", "positionFromSun": 8,
"fractionOfEarthMass": 17.0,
"distanceFromSunInAUs": 30.1, "hasRings": false]


]
```

2. Define a `Planet` `struct` that will be created from the data:

```
struct Planet {
        let name: String
        let positionFromSun: Int
        let fractionOfEarthMass: Double
        let distanceFromSunInAUs: Double
        let hasRings: Bool
}
```

3. Taking this one step at a time, create a function that will take one-planet `dictionaries` and make a `Planet` struct, if it can. We'll use a `guard` statement to ensure that the dictionary has all the values we expect:

```
func makePlanet(fromInput input: [String: Any]) -> Planet? {
        guard let name = input["name"] as? String,
                let positionFromSun = input["positionFromSun"]
as? Int,
                let fractionOfEarthMass =
input["fractionOfEarthMass"] as? Double,
                let distanceFromSunInAUs =
input["distanceFromSunInAUs"] as? Double,
                let hasRings = input["hasRings"] as? Bool else {
                return nil
        }
        return Planet(
                name: name,
                positionFromSun: positionFromSun,
                fractionOfEarthMass: fractionOfEarthMass,
                distanceFromSunInAUs: distanceFromSunInAUs,
                hasRings: hasRings)
}
```

4. Now that we can handle individual planet data, create a function that will take an array of planet dictionaries and make an array of `Planet` structs, using a `guard` statement to ensure that we successfully create a `Planet` struct:

```
func makePlanets(fromInput input: [[String: Any]]) -> [Planet] {
        var planets = [Planet]()
```

```
        for inputItem in input {
                guard let planet = makePlanet(fromInput:
    inputItem) else { continue }
                planets.append(planet)
        }
        return planets
}
```

How it works...

The guard statement works in a very similar way to an if statement, as optional values can be unwrapped and chained together in the same way. Since our planet data contains strings, ints, floats, and Booleans, the dictionary is of the [String: Any] type. So, to create our Planet struct, we will need to check whether the expected values exist for given keys and cast them to the correct type.

In our makePlanet function, we use the guard keyword and then access and conditionally cast all the values we require from the planet data dictionary. If any of these conditional casts fail, the else block, which is defined after the guard statement, is executed. We have defined our function to return an optional Planet, so if we don't have the information expected, the guard will fail, and return nil:

```
func makePlanet(fromInput input: [String: Any]) -> Planet? {
    guard let name = input["name"] as? String,
        let positionFromSun = input["positionFromSun"] as? Int,
        let fractionOfEarthMass = input["fractionOfEarthMass"]
as? Double,
        let distanceFromSunInAUs =
input["distanceFromSunInAUs"] as? Double,
        let hasRings = input["hasRings"] as? Bool else {
        return nil
    }
    return Planet(
        name: name,
        positionFromSun: positionFromSun,
        fractionOfEarthMass: fractionOfEarthMass,
        distanceFromSunInAUs: distanceFromSunInAUs,
        hasRings: hasRings)
}
```

Any value unwrapped by the guard statement is made available to any code below the guard statement in the same scope; this makes the guard statement perfect for ensuring that input values are as expected before continuing. This removes the need to nest our code within an if block. The unwrapped values are then used to initialize the Planet struct.

As we have seen, a guard statement is for breaking execution when the guard condition fails, and therefore, the compiler ensures that an execution breaking statement is placed in the else block; this could be, for example, return, break, or continue.

In the makePlanets function, we use a for loop to iterate through the dictionaries and try to create a Planet struct from each one. If our makePlanet call returns nil, we call continue to skip this iteration of the for loop and jump to the next iteration:

```
func makePlanets(fromInput input: [[String: Any]]) -> [Planet] {
    //...
    for inputItem in input {
        guard let planet = makePlanet(fromInput: inputItem)
else { continue }
        planets.append(planet)
    }
    //...
}
```

There's more...

The makePlanets function accepts an array of planet data dictionaries and returns an array of Planet structs. If the array provided is empty, we may decide that this is not a valid input to our function and want to throw an error; guard can help with this too.

> **Note**
>
> Since our array of planet information happens to resemble our Planet struct, we could utilize the Codable protocol to automatically serialize our data into an instance of Planet:
>
> ```
> extension Planet: Codable {
>     init(dictionary: [String: Any]) throws {
>         self = try JSONDecoder().decode(Planet.self, from:
> JSONSerialization.data(withJSONObject: dictionary))
>     }
> }
> do {
>     let planet = try Planet(dictionary: inputData.first!)
> } catch {
>     print(error)
> }
> ```
>
> We'll learn more about working with JSON and Codable in *Chapter 5*, but it's always fun to learn different ways to control and handle information!

We can check that any conditional statement is `true` with `guard`, and if it isn't, we can throw an error:

```swift
enum CreationError: Error {
    case noData
}

func makePlanets(fromInput input: [[String: Any]]) throws -> [Planet]
{
    guard input.count > 0 else {
        throw CreationError.noData
    }
    //...
}
```

See also

Further information about `guard` statements can be found in Apple's documentation on the Swift language at `http://swiftbook.link/docs/guard`.

Doing it later with defer

Typically, when we call a function, control passes from the call site to the function, and then the statements within the function are executed sequentially until either the end of the function or until a return statement. Control then returns to the call site.

In the following diagram, the print statements are executed in the order *1*, *2*, and *3*:

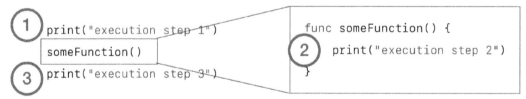

Figure 3.2 – The print statements

Sometimes, it can be useful to execute some code after the function has returned, but before control has been returned to the call site. This is the purpose of Swift's `defer` statement. In the following example, step *3* is executed after step *2*, even though it is defined above it:

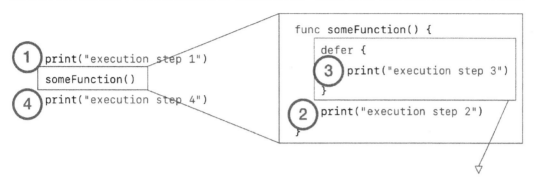

Figure 3.3 – The defer statement

In this recipe, we will explore how to use defer and when it can be helpful.

Getting ready

A defer statement can be useful to change the state once a function's execution is complete or to clean up values that are no longer needed. Let's look at an example of updating the state with a defer statement.

Let's create a new playground for this recipe.

How to do it...

Imagine that we have video game reviews and we want to classify them based on their star rating. Let's see how:

1. Define the options that a video game review may be classified into:

    ```
    enum VideoGameReviewClass {
            case bad
            case average
            case good
            case brilliant
    }
    ```

2. Create an object to do the classification:

    ```
    class VideoGameReviewClassifier {
            func classify(forStarsOutOf10 stars: Int) ->
    VideoGameReviewClass {
                    if stars > 8 {
                            return .brilliant // 9 or 10
                    } else if stars > 6 {
                            return .good // 7 or 8
    ```

```
        } else if stars > 3 {
                return .average // 4, 5 or 6
        } else {
                return .bad // 1, 2 or 3
        }
      }
    }
```

3. Use classifier to classify the review:

```
let classifier = VideoGameReviewClassifier()
let review1 = classifier.classify(forStarsOutOf10: 9)
print(review1) // brilliant
```

4. This works great, but for the purpose of this example, let's imagine that this classification was a long-running process, and we wanted to keep track of the state of the classifier, so we could externally check whether the classifier was in the middle of classifying or was completed. Define the possible classification states:

```
enum ClassificationState {
        case initial
        case classifying
        case complete
}
```

5. Update our class classifier to hold and update the state, using a defer statement to move to the complete state:

```
class VideoGameReviewClassifier {
        var state: ClassificationState = .initial

        func classify(forStarsOutOf10 stars: Int) ->
VideoGameReviewClass {
                state = .classifying
                defer {
                        state = .complete
                }
                if stars > 8 {
                        return .brilliant // 9 or 10
                } else if stars > 6 {
                        return .good // 7 or 8
                } else if stars > 3 {
                        return .average // 4, 5 or 6
                } else {
                        return .bad // 1, 2 or 3
```

```
        }
      }
    }
```

6. Use classifier to classify the review and check the state:

    ```
    let classifier = VideoGameReviewClassifier()
    let review1 = classifier.classify(forStarsOutOf10: 9)
    print(review1) // brilliant
    print(classifier.state) // complete
    ```

How it works...

The `classify` method we defined in the preceding steps takes an input rating and then returns `VideoGameReviewClass`, based on this rating:

```
func classify(forStarsOutOf10 stars: Int) -> VideoGameReviewClass {
    //...
    if stars > 8 {
            return .brilliant // 9 or 10
    } else if stars > 6 {
            return .good // 7 or 8
    } else if stars > 3 {
            return .average // 4, 5 or 6
    } else {
            return .bad // 1, 2 or 3
    }
}
```

While doing that, it also updates a `state` value to indicate where the method is in the classification process:

```
state = .classifying
defer {
    state = .complete
}
```

The `defer` statement allows the state to be updated once the method has returned.

If we were to write this method without the `defer` statement, we would have to transition to the complete state within each branch of the `if` statement before returning a value, as nothing after this will be executed. The end of that method will look as follows:

```
if stars > 8 {
    state = .complete
    return .brilliant // 9 or 10
} else if stars > 6 {
```

```
        state = .complete
        return .good // 7 or 8
} else if stars > 3 {
        state = .complete
        return .average // 4, 5 or 6
} else {
        state = .complete
        return .bad // 1, 2 or 3
}
```

This repetition of updating the state can be avoided when we use the `defer` statement:

```
defer {
        state = .complete
}
```

To defer code, simply use the `defer` keyword, with the code to be deferred defined in curly brackets; this code will be run after the method has returned, but before the control flow is returned to the caller.

There's more...

You can define multiple `defer` statements within a method, and they are executed in the reverse order that they were defined, so the last `defer` statement defined is the first one executed after the method returns.

To demonstrate, add a new state that we'll switch to when completing classifications subsequent to the first:

```
enum ClassificationState {
        case initial
        case classifying
        case complete
        case completeAgain
}
```

Now, let's amend our `classifier` to keep track of the number of classifications it makes, and changes to the `completeAgain` state if more than one classification has been completed:

```
class VideoGameReviewClassifier {
        var state: ClassificationState = .initial
        var numberOfClassifications = 0

        func classify(forStarsOutOf10 stars: Int) ->
    VideoGameReviewClass {
                state = .classifying
```

```
            defer {
                    numberOfClassifications += 1
            }
            defer {
                    if numberOfClassifications > 0 {
                            state = .completeAgain
                    } else {
                            state = .complete
                    }
            }

            if stars > 8 {
                    return .brilliant // 9 or 10
            } else if stars > 6 {
                    return .good // 7 or 8
            } else if stars > 3 {
                    return .average // 4, 5 or 6
            } else {
                    return .bad // 1, 2 or 3
            }
        }
}
```

Now, we change how we use `classifier`; the second time we use it, it will complete with a different state:

```
let classifier = VideoGameReviewClassifier()
let review1 = classifier.classify(forStarsOutOf10: 9)
print(review1) // brilliant
print(classifier.state) // complete
print(classifier.numberOfClassifications) // 1

let review2 = classifier.classify(forStarsOutOf10: 2)
print(review2) // bad
print(classifier.state) // completeAgain
print(classifier.numberOfClassifications) // 2
```

Since we have now defined two `defer` statements, let's take another look to understand the order in which they are executed:

```
defer {
        numberOfClassifications += 1
}
defer {
```

```
        if numberOfClassifications > 0 {
                state = .completeAgain
        } else {
                state = .complete
        }
    }
```

As discussed earlier, the last defined `defer` statement is executed first. So, on the first classification, once the method returns, the last `defer` statement is executed, and the state is changed to `complete` because `numberOfClassifications` will be 0. Then, the first `defer` statement is executed, adding 1 to the `numberOfClassifications` variable, which will now be 1.

On the second classification, once the method returns, the last `defer` statement will execute and change the state to `completeAgain`, since `numberOfClassifications` is greater than 0. Finally, the first `defer` statement will execute, incrementing `numberOfClassifications` and making it 2.

If the `defer` statements had been the other way around, the state would always change to `completeAgain`, as `numberOfClassifications` would have incremented to 1 before the check was made.

See also

Further information about `defer` statements can be found in Apple's documentation on the Swift language at `http://swiftbook.link/docs/defer`.

Bailing out with fatalError and precondition

It's comforting to think that in the code you write, everything will always happen as expected, and your program can handle any eventuality. However, sometimes things can go wrong – really wrong. A situation could arise that you know is possible but don't expect to ever happen, and the program should terminate if it does. In this recipe, we will look at two issues like this – `fatalError` and `precondition`.

Getting ready

Let's reuse our example from the previous recipe – we have an object that can be used to classify video game reviews, based on how many stars out of 10 the review gave the video game. However, let's simplify its use, and say that we only intend for a classifier object to classify one, and only one, video game review.

How to do it...

Let's set up our video game classifier to only be used once, and only accept ratings out of 10:

1. Define the classification state and the video game review class:

```
enum ClassificationState {
        case initial
        case classifying
        case complete
}


enum VideoGameReviewClass {
        case bad
        case average
        case good
        case brilliant
}
```

2. Redefine our classifier object, using `precondition` and `fatalError` to indicate situations that are not expected to occur and would cause a problem:

```
class VideoGameReviewClassifier {
        var state: ClassificationState = .initial

        func classify(forStarsOutOf10 stars: Int) ->
VideoGameReviewClass {
                precondition(state == .initial, "Classifier state
must be initial")
                state = .classifying
                defer {
                        state = .complete
                }
                if stars > 8 && stars <= 10 {
                        return .brilliant // 9 or 10
                } else if stars > 6 {
                        return .good // 7 or 8
                } else if stars > 3 {
                        return .average // 4, 5 or 6
                } else if stars > 0 {
                        return .bad // 1, 2 or 3
                } else {
                        fatalError("Star rating must be between 1
and 10")
                }
        }
}
```

```
    }

    let classifier = VideoGameReviewClassifier()
    let review1 = classifier.classify(forStarsOutOf10: 9)
    print(review1) // brilliant
    print(classifier.state) // complete
```

How it works...

We only want to use the classifier once; therefore, when we begin to classify a video game review, the current state should be `initial`, as this object has never been classified before and shouldn't be in the middle of classifying. If that is not the case, the classifier is being used incorrectly, and we should terminate the execution of the code:

```
func classify(forStarsOutOf10 stars: Int) -> VideoGameReviewClass {
        precondition(state == .initial, "Classifier state must be
initial")
        //...
}
```

We state a precondition using the `precondition` keyword, and we provide a Boolean statement that we expect to be `true` and an optional message. If this Boolean statement is not `true`, the execution of the code will terminate, and the message will be displayed in the console.

In our example, we are making it a precondition that the state must be `initial` when calling this method. When our classifier performs the classification, it expects a number of stars between 1 and 10. However, the method accepts `Int` as a parameter, so any integer value can be provided, positive or negative.

If the value provided is not between 1 and 10, and the classifier cannot provide a valid `VideoGameReviewClass`, then the classifier is being used incorrectly, and we should terminate the execution of the code:

```
func classify(forStarsOutOf10 stars: Int) -> VideoGameReviewClass {
        //...
        if stars > 8 && stars <= 10 {
                return .brilliant // 9 or 10
        } else if stars > 6 {
                return .good // 7 or 8
        } else if stars > 3 {
                return .average // 4, 5 or 6
        } else if stars > 0 {
                return .bad // 1, 2 or 3
        } else {
```

```
        fatalError("Star rating must be between 1 and 10")
    }
}
```

The if-else statement covers all the valid `VideoGameReviewClass` options for the provided stars, so if none of these are triggered, we use a fatal error to indicate incorrect usage. This is done using the `fatalError` keyword, and an optional message can be provided.

See also

Further information about `fatalError` can be found in Apple's documentation on the Swift language at `http://swiftbook.link/docs/fatalerror`.

4

Generics, Operators, and Nested Types

Swift provides a number of advanced features for building functionality that is flexible but well-defined so that it feels like you are extending the language itself. In this chapter, we will examine two of these features: **generics** and **operators**. We will also see how **nested types** allow **logical grouping**, **access control**, and **namespacing** for your constructs.

By the end of this chapter, you will be empowered with advanced tooling that can help you achieve some neat goals with your Swift code!

In this chapter, we will cover the following recipes:

- Using generics with types
- Using generics with functions
- Using generics with protocols
- Using advanced operators
- Defining option sets
- Creating custom operators
- Nesting types and namespacing

Technical requirements

All the code for this chapter can be found in this book's GitHub repository at `https://github.com/PacktPublishing/Swift-Cookbook-Third-Edition/tree/main/Chapter%204`.

Using generics with types

When we build things in Swift that interact with other types, we often specify the type we are interacting with directly. This is helpful because it means we know the capabilities that the type has. We can put those capabilities to use and ensure that the outputs have the correct type. However, we then have a construct that can only interact with the specified type; it can't be reused with other types, even if the concepts are the same.

Generics give us the advantage of having a defined type while being generically applicable to other types. It is, perhaps, best illustrated with an example.

In this recipe, we will create a generic class that stores the last five things it was given and returns them all upon request.

Getting ready

In this recipe, we will create a custom collection object that will store the last five strings that the user copied so that they can paste not just the last string copied, but any of the last five. You can add strings to the list and ask for all the strings in the list, which will be returned from newest to oldest.

Add the following code to a new playground:

```
class RecentList {
    var slot1: String?
    var slot2: String?
    var slot3: String?
    var slot4: String?
    var slot5: String?

    func add(recent: String) {
        // Move each slot down 1
        slot5 = slot4
        slot4 = slot3
        slot3 = slot2
        slot2 = slot1
        slot1 = recent
    }

    func getAll() -> [String] {
        var recent = [String]()
        if let slot1 = slot1 {
            recent.append(slot1)
        }
        if let slot2 = slot2 {
            recent.append(slot2)
```

```
        }
        if let slot3 = slot3 {
                recent.append(slot3)
        }
        if let slot4 = slot4 {
                recent.append(slot4)
        }
        if let slot5 = slot5 {
                recent.append(slot5)
        }
        return recent
    }
}


let recentlyCopiedList = RecentList()
recentlyCopiedList.add(recent: "First")
recentlyCopiedList.add(recent: "Next")
recentlyCopiedList.add(recent: "Last")
var recentlyCopied = recentlyCopiedList.getAll()
print(recentlyCopied) // Last, Next, First
```

This is great – it does just what we want. Now, let's say that we want to add a list of five recent contacts to a contacts app. The concept is exactly the same as the list of copied strings, as we want to do the following:

- Add something to a list
- Get all the things on the list so that we can present them to the user

However, because we specified that the `RecentList` object can only work with strings, it can't work with my custom `Person` object. We can use generics to make this more useful.

Let's see how to do this by making `RecentList` use generics.

How to do it...

We will update our `RecentList` code to use generics, so it can be used with other types:

1. Amend the `RecentList` object to define a generic type, `ListItemType`, which we use in place of `String`:

    ```
    class RecentListGeneric<ListItemType> {
            var slot1: ListItemType?
            var slot2: ListItemType?
            var slot3: ListItemType?
    ```

```
        var slot4: ListItemType?
        var slot5: ListItemType?

        func add(recent: ListItemType) {
                // Move each slot down 1
                slot5 = slot4
                slot4 = slot3
                slot3 = slot2
                slot2 = slot1
                slot1 = recent
        }

        func getAll() -> [ListItemType] {
                var recent = [ListItemType]()
                if let slot1 = slot1 {
                        recent.append(slot1)
                }
                if let slot2 = slot2 {
                        recent.append(slot2)
                }
                if let slot3 = slot3 {
                        recent.append(slot3)
                }
                if let slot4 = slot4 {
                        recent.append(slot4)
                }
                if let slot5 = slot5 {
                        recent.append(slot5)
                }
                return recent
        }
}
```

2. Provide a specified type, String, when creating RecentList, which will be used to replace the generic type for this instance of RecentList:

```
let recentlyUsedWordList = RecentListGeneric<String>()
recentlyUsedWordList.add(recent: "First")
recentlyUsedWordList.add(recent: "Next")
recentlyUsedWordList.add(recent: "Last")
var recentlyUsedWords = recentlyUsedWordList.getAll()
print(recentlyUsedWords) // Last, Next, First
```

> **Note**
>
> Instead of using generics, we could have replaced all the `String` references in `RecentList` with Any, which would allow it to accept any type. However, this would allow the list to be made up of different types of things, which is not what we want. It would also require us to cast values that are returned, to make them useful.

Let's examine how our newly genericized `RecentList` can be used for the other example we discussed earlier, the list of recent contacts:

1. Create a simple `Person` object:

```
class Person {
    let name: String
    init(name: String) {
        self.name = name
    }
}
```

2. Create some people to add to our recent contact list:

```
let rod = Person(name: "Rod")
let jane = Person(name: "Jane")
let freddy = Person(name: "Freddy")
```

3. Create a new `RecentList` object, providing the specific `Person` type:

```
let lastCalledList = RecentList<Person>()
```

4. Add person objects to this list:

```
lastCalledList.add(recent: freddy)
lastCalledList.add(recent: jane)
lastCalledList.add(recent: rod)
```

5. Get all the people in the list, and since this is typed as an array of `Person` objects, print their name property:

```
let lastCalled = lastCalledList.getAll()
for person in lastCalled {
    print(person.name)
}
// Rod
// Jane
// Freddy
```

We now have a generic `RecentList` class that we have used with both strings and a custom `Person` class.

How it works...

To add generics to a class or struct, the generic type is defined in angle brackets after the class or struct name, and can be given any type name, although it should begin with a capital letter like other type names:

```
class RecentList<ListItemType> {
    //...
}
```

This generic type now becomes a stand-in for the specific type that will be specified when it is used, and we can use this wherever we would use the specific type.

It can be used as a property type:

```
var slot1: ListItemType?
```

It can be used as a parameter value:

```
func add(recent: ListItemType)
```

And it can be used as a return type:

```
func getAll() -> [ListItemType]
```

In many other programming languages that have a generics system, the generic type is often given a one-letter type name, usually T. Swift aims to be concise, but not at the expense of clarity, so I suggest using a more descriptive type name.

A descriptive type name becomes especially important if you have multiple generic types, which you can have as a comma-separated list within the angle brackets:

```
class RecentList<ListItemType, SomeOtherType> {
    //...
}
```

We have now created a generic RecentList object that can be used with any type.

There's more...

While being extremely generic has its advantages, you may wish to constrain which types can be used for the generic type, especially if you need to use some features of that constrained type.

Let's say that in addition to returning an array of items from RecentList, we want to be able to print out the list directly. To do this, we need to ensure that the type of item used in RecentList is something that can be converted into a string to be printed. There is already a CustomStringConvertible protocol that defines this behavior, so we want to ensure that any specific type used with `RecentList` conforms to `CustomStringConvertible`:

```
class RecentList<ListItemType: CustomStringConvertible> {
    //...
}
```

We add the constraint after the generic type name, separated by a colon, similar to how we specify protocol conformance and class inheritance. Indeed, while this example constrains the generic type to implement a protocol, we can instead specify a class that the specific type must be, or inherit from.

Now that we have this constraint, we can be sure that any specific type given will conform to `CustomStringConvertible`, and will therefore have a description string that we can print, so let's create a method to do that:

```
class RecentList<ListItemType: CustomStringConvertible> {
    func printRecentList() {
        for item in getAll() {
            let printableItem = String(describing: item)
            print(printableItem)
        }
    }
}
//...
```

The only thing left to do is to make our `Person` class conform to `CustomStringConvertible` so that it can continue to be used as a specific type in `RecentList`:

```
extension PersonPrintable: CustomStringConvertible {
    public var description: String {
        return name
    }
}
```

Now we can use this functionality with our `String` type's `RecentList` and our `Person` type's `RecentList`:

```
// Using Strings type
let recentlyUsedWordList = RecentListPrintable<String>()
recentlyUsedWordList.add(recent: "First")
recentlyUsedWordList.add(recent: "Next")
recentlyUsedWordList.add(recent: "Last")
recentlyUsedWordList.printRecentListPrintable()
// Last
// Next
// First
```

```
// Using PersonPrintable type
let rod = PersonPrintable(name: "Rod")
let jane = PersonPrintable(name: "Jane")
let freddy = PersonPrintable(name: "Freddy")

let lastCalledList = RecentListPrintable<PersonPrintable>()
lastCalledList.add(recent: freddy)
lastCalledList.add(recent: jane)
lastCalledList.add(recent: rod)
lastCalledList.printRecentListPrintable()
// Rod
// Jane
// Freddy
```

By constraining the generic type, we could use features that we knew the type would have, to provide additional functionality.

See also

Further information about generic types can be found in Apple's documentation on the Swift language at http://swiftbook.link/docs/generics.

Using generics with functions

In addition to being able to specify generic types, you can use generics to build functions that are both widely applicable and strongly typed. In this recipe, we will use generics with functions.

Getting ready

In this recipe, we won't be using any components from the previous recipes, so you can create a new playground for this recipe.

How to do it...

We will use generics to create a function to help with placing values into a dictionary:

1. Create a generic function that inserts the same value into a dictionary for multiple keys:

    ```
    func makeDuplicates<ItemType>(of item: ItemType, withKeys keys:
    Set<String>) -> [String: ItemType] {
        var duplicates = [String: ItemType]()
        for key in keys {
            duplicates[key] = item
        }
    ```

```
          return duplicates
     }
```

2. Use this function, passing in a single value and multiple keys, and the value is stored against each of the given keys:

```
let awards: Set<String> = ["Best Visual Effects",
                           "Best Cinematography",
                           "Best Original Score",
                           "Best Film Editing"]

let oscars2022 = makeDuplicates(of: "Dune", withKeys: awards)
print(oscars2022["Best Visual Effects"] ?? "")
// Dune
print(oscars2022["Best Cinematography"] ?? "")
// Dune
```

How it works...

Just like generics for types, the generic type for a function is specified within angle brackets:

```
func makeDuplicates<ItemType>(of item: ItemType, withKeys keys:
Set<String>) -> [String: ItemType] {
    //...
}
```

The defined generic type name can then be used as a type definition within the rest of the function definition. In our example, we want to define the type of our input item to be duplicated, and we also want the values that are held in the dictionary to be returned to be of the same type.

Instead of using generics, we could have used the Any type in place of the generic type:

```
func makeDuplicates(of item: Any, withKeys keys: Set<String>) ->
[String: Any] {
    //...
}
```

However, this approach presents a few problems for anyone using this function:

- They will get back a dictionary containing values of the Any type, which will need to be cast to a more useful type.

- Without seeing the implementation, they can't be sure that the dictionary contains values of the same type. One key may have a String stored against it, and another may have an Int.

- Without seeing the implementation, they can't be sure that the values of the returned dictionary are of the same type as the item provided.

By using a generic type, we allow the functionality to be widely applicable while enforcing our type logic at compile time.

You'll notice that unlike instantiating a type with generics, we don't need to explicitly state the specific type to use when executing the function:

```
let oscars2022 = makeDuplicates(of: "Dune", withKeys: awards)
```

This is because the compiler is able to infer it from the type of the first parameter provided. Since Dune is a string, and the compiler knows that the parameter has the ItemType generic type, the compiler infers that for this use of the method, the ItemType generic type becomes the specific type of String.

There's more...

We can increase the usability of our function by providing a generic type for the set of keys we provide as the second parameter:

```
func makeDuplicates<ItemType, KeyType>(of item: ItemType, withKeys
keys: Set<KeyType>) -> [KeyType: ItemType] {
    var duplicates = [KeyType: ItemType]()
    for key in keys {
        duplicates[key] = item
    }
    return duplicates
}
```

Multiple generic types are defined just as they were in the previous recipe, as a comma-separated list within angle brackets.

All the collection types in Swift (array, dictionary, set, and so on) use generics, and in the preceding function, we are passing the generic type from our function into the set.

Therefore, KeyType must conform to Hashable, since this is required for use in a set.

If we wanted to make this constraint explicit or constrain the generic type for some other reason, this is defined after a colon:

```
func makeDuplicates<ItemType, KeyType: Hashable>(of item: ItemType,
withKeys keys: Set<KeyType>) -> [KeyType: ItemType] {
    var duplicates = [KeyType: ItemType]()
    for key in keys {
        duplicates[key] = item
    }
    return duplicates
}
```

Just as with the previous example, if both specific types we are using can be inferred from the input or output, we don't need to specify them:

```
let awards: Set<String> = ["Best Visual Effects",
                                   "Best Cinematography",
                                   "Best Original Score",
                                   "Best Film Editing"]
let oscars2022Generic = makeDuplicatesGeneric(of: "Dune", withKeys:
awards)
print(oscars2022Generic["Best Visual Effects"] ?? "")
print(oscars2022Generic["Best Cinematography"] ?? "")
```

We have used two generic types to improve the flexibility of our function.

See also

Further information about generic functions can be found in Apple's documentation on the Swift language at https://docs.swift.org/swift-book/documentation/the-swift-programming-language/generics/#Generic-Functions.

Using generics with protocols

So far in this chapter, we have seen how to use generics within types and functions. In this recipe, we will round off our journey through generics in Swift by looking at how they can be used in protocols. This will allow us to produce abstract interfaces while maintaining strongly typed requirements that allow for a more descriptive model.

In this recipe, we will build a model for a transport app in the UK with the goal of providing the distance and duration of a journey for different methods of transport.

Getting ready

In this recipe, we won't be using any components from the previous recipes, so you can create a new playground for this recipe.

How to do it...

The ways that people travel are very different, so let's start by defining transport methods in a generic way, and then specify what those travel methods are:

1. Define a protocol to define the features of a transport method:

```
protocol TransportMethod {
        associatedtype CollectionPoint: TransportLocation
```

```
        var defaultCollectionPoint: CollectionPoint { get }
        var averageSpeedInKPH: Double { get }
}
```

2. Create a struct for traveling by train that implements the `TransportMethod` protocol:

```
struct Train: TransportMethod {
        typealias CollectionPoint = TrainStation

        // User's home or nearest station
        var defaultCollectionPoint: CollectionPoint {
                return TrainStation.BMS
        }

        var averageSpeedInKPH: Double {
                return 100
        }
}
```

3. We need to define the `TrainStation` type that we put as `CollectionPoint`. Let's do that as an enum:

```
enum TrainStation: String, TransportLocation {
        case BMS = "Bromley South"
        case VIC = "London Victoria"
        case RAI = "Rainham (Kent)"
        case BTN = "Brighton (East Sussex)"
}
```

4. Since we plan to calculate the distance and duration of a journey, let's create a `Journey` object to represent that journey from a starting point to an endpoint:

```
class Journey<TransportType: TransportMethod> {
        let start: TransportType.CollectionPoint
        let end: TransportType.CollectionPoint

        init(start: TransportType.CollectionPoint,
                end: TransportType.CollectionPoint) {
                self.start = start
                self.end = end
        }
}
```

5. Add the transport method as a property of the journey as this will be used for the duration calculation:

```
class Journey<TransportType: TransportMethod> {
    let start: TransportType.CollectionPoint
    let end: TransportType.CollectionPoint
    let method: TransportType

    init(method: TransportType,
         start: TransportType.CollectionPoint,
         end: TransportType.CollectionPoint) {
        self.start = start
        self.end = end
        self.method = method
    }
}
```

6. To calculate the distance of our journey, we need the start and end to have definite locations. So, define a protocol that provides these locations:

```
protocol TransportLocation {
    var location: CLLocation { get }
}
```

7. Import the `CoreLocation` framework at the top of the playground:

```
import CoreLocation
```

8. Constrain the `CollectionPoint` associated type on `TransportMethod`, so that it must conform to the `TransportLocation` protocol we have just created:

```
protocol TransportMethod {
    associatedtype CollectionPoint: TransportLocation
    var defaultCollectionPoint: CollectionPoint { get }
    var averageSpeedInKPH: Double { get }
}
```

9. Use the locations of the start and end of the `CollectionPoint` object to calculate the distance and duration of the journey:

```
class Journey<TransportType: TransportMethod> {
    var start: TransportType.CollectionPoint
    var end: TransportType.CollectionPoint

    let method: TransportType
    var distanceInKMs: Double
    var durationInHours: Double
```

```
        init(method: TransportType, start: TransportType.
CollectionPoint, end: TransportType.CollectionPoint) {
            self.start = start
            self.end = end
            self.method = method
            // CoreLocation provides the distance in meters, so we
divide by 1000 to get kilometers
            distanceInKMs = end.location.distance(from: start.
location) / 1000
            durationInHours = distanceInKMs / method.
averageSpeedInKPH
        }
}
```

10. Ensure our `TrainStation` enum conforms to `TransportLocation`, which is now a requirement:

```
enum TrainStation: String, TransportLocation {
    case BMS = "Bromley South"
    case VIC = "London Victoria"
    case RAI = "Rainham (Kent)"
    case BTN = "Brighton (East Sussex)"
    // Full list of UK train stations codes can be found at
    // http://www.nationalrail.co.uk/static/documents/content/
station_codes.csv

    var location: CLLocation {
        switch self {
        case .BMS: return CLLocation(latitude: 51.4000504,
longitude: 0.0174237)
        case .VIC: return CLLocation(latitude: 51.4952103,
longitude: -0.1438979)
        case .RAI: return CLLocation(latitude: 51.3663,
longitude: 0.61137)
        case .BTN: return CLLocation(latitude: 50.829,
longitude: -0.14125)
        }
    }
}
```

11. Use our `Journey` object to calculate the distance and duration of a train journey:

```
let trainJourney = Journey(method: Train(), start: TrainStation.
BMS, end: TrainStation.VIC)
let distanceByTrain = trainJourney.distanceInKMs
```

```
let durationByTrain = trainJourney.durationInHours
print("Journey distance: \(distanceByTrain) km")
print("Journey duration: \(durationByTrain) hours")
```

How it works...

At the outset, it may not be clear which is the best structure to use to define a transport method, and there might be different structures appropriate for different travel methods. Therefore, we can define a transport method as a protocol that appropriate types can conform to:

```
protocol TransportMethod { associatedtype CollectionPoint
    var defaultCollectionPoint: CollectionPoint { get }
    var averageSpeedInKPH: Double { get }
}
```

We define an associated generic type that we name `CollectionPoint`, which will represent the type of location that someone can be collected from when using this transport method. By using generics, we have ultimate flexibility in how a transport method chooses to define what can serve as a collection point.

Having defined an associated type, it can then be used as a placeholder in properties and methods for the specific type that will be defined when the protocol is used. We use it to define a default collection point that each transport method should provide.

Each transport method also provides an average speed, which will be used later in calculating the travel time.

Let's look at a concrete example of a transport method to help define the model further:

```
struct Train: TransportMethod {
    typealias CollectionPoint = TrainStation
    // User's home or nearest station
    var defaultCollectionPoint: TrainStationPoint {
        return TrainStation.BMS
    }

    var averageSpeedInKPH: Double {
        return 100
    }
}
```

For `Train` to conform to the `TransportMethod` protocol, we must provide a specific version of the `CollectionPoint` generic type that is required by the protocol. In the case of traveling by train, the collection point will be a train station, so we now have to define the `TrainStation` type:

```
enum TrainStation: String {
    case BMS = "Bromley South"
    case VIC = "London Victoria"
    case RAI = "Rainham (Kent)"
    case BTN = "Brighton (East Sussex)"
    // Full list of UK train stations codes at
    // http://www.railwaycodes.org.uk/stations/station1.shtm
}
```

Since there are a finite number of train stations that are discretely definable, an enum is a good way to represent them. I've only listed a small number in the preceding code block, for brevity.

Our goal is to model a journey and calculate the duration of the journey over specific transport methods, so let's create a `Journey` object:

```
class Journey<TransportType: TransportMethod> {
    let start: TransportType.CollectionPoint
    let end: TransportType.CollectionPoint

    init(start: TransportType.CollectionPoint, end: TransportType.
CollectionPoint) {
        self.start = start
        self.end = end
    }
}
```

A journey takes place from one point to another, so we take the journey's start and end as input parameters. We need to have the flexibility to provide any type as the start and end, but we need them to be types connected to a transport method, with the same type for the start and end values. To accomplish this, we can have a generic type constrained to conform to the `TransportMethod` protocol. We can then define our start and end property types by referencing the `CollectionPoint` associated type of the generic type.

Our goal is to calculate the duration of a journey. To do this, we will need the speed of travel during the journey and the distance from start to end. Our `TransportMethod` protocol defines that it will provide an average speed, so let's also take the transport method as an input to our journey:

```
class Journey<TransportType: TransportMethod> {
    let start: TransportType.CollectionPoint
    let end: TransportType.CollectionPoint
    let method: TransportType
```

```
    init(method: TransportType, start: TransportType.CollectionPoint,
end: TransportType.CollectionPoint) {
        self.start = start
        self.end = end
        self.method = method
    }
}
```

To get the distance of the journey, we need to calculate the distance between the start and end, but the type of both the start and end of the journey is the generic `CollectionPoint` type, which could be any type, and so does not have any location information that we can use to calculate the distance.

To solve this, let's constrain `CollectionPoint` so that it must conform to a new protocol, `TransportLocation`:

```
protocol TransportLocation {
    var location: CLLocation { get }
}
```

Anything conforming to `TransportLocation` must provide a location in the form of a `CLLocation` object. The `CLLocation` object is part of the `CoreLocation` framework on iOS. Further investigation of the `CoreLocation` framework is outside the scope of this book, but it's enough to know that it provides ways to calculate the distance between two `CLLocation` objects, and we need to include the following at the top of this playground to use it:

```
import CoreLocation
```

With our `TransportLocation` protocol defined, we can constrain the `CollectionPoint` associated type on the `TransportMethod` protocol:

```
protocol TransportMethod {
    associatedtype CollectionPoint: TransportLocation
    var defaultCollectionPoint: CollectionPoint { get }
    var averageSpeedInKPH: Double { get }
}
```

Since our `CollectionPoint` will now conform to `TransportLocation`, and therefore must have a location property, we can go back to our `Journey` object and use this to calculate the distance of the journey and the duration:

```
class Journey<TransportType: TransportMethod> {
    var start: TransportType.CollectionPoint
    var end: TransportType.CollectionPoint
    let method: TransportType
    var distanceInKMs: Double
```

```
      var durationInHours: Double

    init(method: TransportType, start: TransportType.CollectionPoint,
end: TransportType.CollectionPoint) {
         self.start = start
         self.end = end
         self.method = method
         // CoreLocation provides the distance in meters, so we
divide by 1000 to get kilometers
         distanceInKMs = end.location.distance(from: start.location)
/ 1000
         durationInHours = distanceInKMs / method.averageSpeedInKPH
    }
}
```

The last thing we need to do is to ensure that our `TrainStation` enum conforms to `TransportLocation` as this is now a requirement. To do this, we just need to declare conformance and add a location property:

```
enum TrainStation: String, TransportLocation {
    case BMS = "Bromley South"
    case VIC = "London Victoria"
    case RAI = "Rainham (Kent)"
    case BTN = "Brighton (East Sussex)"
    // Full list of UK train stations codes can be found at
    // http://www.railwaycodes.org.uk/stations/station1.shtm

    var location: CLLocation {
         switch self {
         case .BMS:
             return CLLocation(latitude: 51.4000504, longitude:
0.0174237)
         case .VIC:
             return CLLocation(latitude: 51.4952103, longitude:
-0.1438979)
         case .RAI:
             return CLLocation(latitude: 51.3663, longitude:
0.61137)
         case .BTN:
             return CLLocation(latitude: 50.829, longitude:
-0.14125)
         }
    }
}
```

Let's see how we would use our travel model to create a journey with specific types:

```
let trainJourney = Journey(method: Train(), start: TrainStation.BMS,
end: TrainStation.VIC)
let distanceByTrain = trainJourney.distanceInKMs
let durationByTrain = trainJourney.durationInHours
print("Journey distance: \(distanceByTrain) km")
print("Journey duration: \(durationByTrain) hours")
```

We have used generics with protocols to create a generic system without prescribing the type of Swift construct we need to use.

There's more...

In this recipe, we made one type conform to `TransportMethod` – this was our `Train` struct. Let's look at another to see how tackling things in a protocol-oriented way allows flexibility in implementation.

In the next `TransportMethod`, we will implement `Road`, but there are a number of different vehicle types that we can use to travel by road, and they may have different average speeds.

1. Since we have a finite list of options for travel by road, let's define it using an `enum`:

    ```
    enum Road: TransportMethod {
        typealias CollectionPoint = CLLocation
        case car
        case motobike case van
        case hgv

        // The users home or current location
        var defaultCollectionPoint: CLLocation {
            return CLLocation(latitude: 51.1, longitude: 0.1)
        }

        var averageSpeedInKPH: Double {
            switch self {
            case .car: return 60
            case .motobike: return 70
            case .van: return 55
            case .hgv: return 50
            }
        }
    }
    ```

2. A journey by train has a finite list of collection points, which is the train stations, but almost anywhere can be a collection point when traveling by road. Therefore, we can define the collection point for Road to be any CLLocation, but CLLocation doesn't conform to TransportLocation. We can solve this by extending CLLocation to add conformance:

```swift
extension CLLocation: TransportLocation {
    var location: CLLocation {
        return self
    }
}
```

Now, we can define a journey by road and calculate the duration:

```swift
let start = CLLocation(latitude: 51.3994669, longitude: 0.0116888)
let end = CLLocation(latitude: 51.2968654, longitude: 0.5053609)
let roadJourney = Journey(method: Road.car, start: start, end: end)
let distanceByRoad = roadJourney.distanceInKMs
let durationByRoad = roadJourney.durationInHours
print("Journey distance: \(distanceByRoad) km")
print("Journey duration: \(durationByRoad) hours")
```

By taking a protocol-orientated approach to tackling the task of calculating a journey's duration, and by using protocol generics, we were able to use completely different, but appropriate, implementations for two transport methods while providing an interface so that they can be handled in a common way.

For a train journey, we used an enum to model the train stations and a struct to model the transport method, and for a road journey, we implemented an enum for the transport method and used the CLLocation object for the transport location.

See also

Further information about associated types can be found in Apple's documentation on the Swift language at http://swiftbook.link/docs/associated-types.

Using advanced operators

Swift is a programming language that takes a relatively small number of well-defined principles and builds on them to create expressive and powerful language features. The concept of mathematical operators, such as +, -, *, and / for addition, subtraction, multiplication, and division, respectively, seems so fundamental as to not warrant a mention. However, in Swift, this common mathematical functionality is built on top of an underlying operator system that is extensible and powerful.

In this recipe, we will look at some of the more advanced operators provided by the Swift standard library, and in the next recipe, we will create our own custom operators.

Getting ready

In this recipe, we won't be using any components from the previous recipes, so you can create a new playground for this recipe.

How to do it...

The operators we will explore are known as bitwise operators and are used to manipulate numerical bit representations.

An integer value in Swift can be represented in its binary form by prefixing the integer literal with 0b:

```
let zero: Int = 0b000
let one: Int = 0b001
let two: Int = 0b010
let three: Int = 0b011
let four: Int = 0b100
let five: Int = 0b101
let six: Int = 0b110
let seven: Int = 0b111
```

A bit is the smallest value in a computer system, consisting of either a 1 or 0. The integers mentioned here can be represented by three bits, which are clearly visible when represented in binary form, as illustrated in the preceding snippet. The integer six can be represented by the three bits 1, 1, and 0.

These binary representations are really useful when you need to represent multiple options in one value. For example, let's say that we want to indicate which devices are supported for a specific feature of an app. The available devices are as listed:

- Phone
- Tablet
- Watch
- Laptop
- Desktop
- TV
- Brain implant

Certain features may be appropriate for all the devices; you may still be working on a feature, which means it isn't currently appropriate for any device; or it may be appropriate for a combination of different devices. We can have Boolean values for each of the devices to indicate whether the feature is supported for that device, but this is not the best solution as there is nothing intrinsically tying the properties to each other, and you could forget to update some of the values as circumstances change.

Instead, we can represent all the supported devices with one integer value, and use each bit of the integer to represent a different device:

```
let phone: Int = 0b0000001
let tablet: Int = 0b0000010
let watch: Int = 0b0000100
let laptop: Int = 0b0001000
let desktop: Int = 0b0010000
let tv: Int = 0b0100000
let brainImplant: Int = 0b1000000
```

To see how this enables us to store multiple devices in one value, let's add together a number of device values:

```
phone = 0b0000001
tablet = 0b0000010
tv = 0b0100000
phone + tablet + tv = 0b0100011
```

As each device is represented by a different bit, the device values are combined by adding the values, and they don't overlap.

To test whether a particular device or combination of devices is supported, we can use a bitwise AND operation. A bitwise AND operation will compare the corresponding bits for two different binary values and will set that bit to 1 in a new binary value if both bit input values are 1. As an example, let's test whether phones are supported in the combined value we created earlier:

Supported Devices	=	0b	0	1	0	0	0	1	1
Phone	=	0b	0	0	0	0	0	0	1
AND Operation Result	=	0b	0	0	0	0	0	0	1

The result only has a 1-bit value for the rightmost bit because this is the only bit that was set to 1 in both the **Supported Devices** value and the **Phone** value.

Once we have that result, we can directly compare it to the value for **Phone**, and if they are equal, then we know that the value of **Supported Devices** included the **Phone** value:

- AND Operation Result = 0b 0 0 0 0 0 0 1

- Phone = 0b 0 0 0 0 0 0 1

We now have a way to combine the possible options into one value, and a way to compare those values to see whether one is contained in another, using bitwise operations. The Swift standard library contains bitwise operators that allow us to perform these operations as easily as other mathematical operations, such as +, -, *, and /.

Typically, an operator will be in the following form:

```
<#left hand side value#> <#operator#> <#right hand side value#>
```

As it is when adding two numbers together, for example:

```
2 + 3
```

In the preceding example, we have these:

- 2: This is the left-hand side value

- 3: This is the right-hand side value

The bit shift operator (<<) will take an integer value on the left-hand side and shift it by the number of bit positions to the right-hand side. Therefore, we can use this to express our intention better when declaring device values:

let	phone: Int	=	1	<<	0	//	0b0000001
let	tablet: Int	=	1	<<	1	//	0b0000010
let	watch: Int	=	1	<<	2	//	0b0000100
let	laptop: Int	=	1	<<	3	//	0b0001000
let	desktop: Int	=	1	<<	4	//	0b0010000
let	tv: Int	=	1	<<	5	//	0b0100000
let	brainImplant:	Int =	1	<<	6	//	0b1000000

The bitwise AND operator (&) will perform the same bit comparison that was previously described manually, and we can use this to create a function to determine whether a particular device exists within the value for the supported devices:

```
var supportedDevices = phone + tablet + tv

func isSupported(device: Int) -> Bool {
    let bitWiseANDResult = supportedDevices & device
    let containsDevice = bitWiseANDResult == device
    return containsDevice
}

let phoneSupported = isSupported(device: phone)
print(phoneSupported) // true

let brainImplantSupported = isSupported(device: brainImplant)
print(brainImplantSupported) // false
```

The Swift standard library also provides operators for the following logical operations:

- **OR**: The OR operation, denoted by |, compares bits and sets the corresponding bit to 1 if either value has the bit set to 1. For our devices, this will mean creating a union between two device combinations:

```
let deviceThatSupportUIKit = phone + tablet + tv
let stationaryDevices = desktop + tv
let stationaryOrUIKitDevices = deviceThatSupportUIKit |
stationaryDevices
let orIsUnion = stationaryOrUIKitDevices == (phone + tablet + tv
+ desktop)
print(orIsUnion) // true
```

- **XOR** (exclusive or): The XOR operation, denoted by ^, will only set the bit to 1 if either value has the bit set to 1, but not if they both do:

```
let onlyStationaryOrUIKitDevices = deviceThatSupportUIKit ^
stationaryDevices
let xorIsUnionMinusIntersection = onlyStationaryOrUIKitDevices
== (phone + tablet + desktop)
print(xorIsUnionMinusIntersection) // true
```

We have seen some of the advanced operators available to us, provided by the Swift standard library.

See also

Further information about advanced operators can be found in Apple's documentation on the Swift language at `http://swiftbook.link/docs/advanced-operators`.

Defining option sets

The use of bitwise operations to hold multiple options in one value is a common pattern and is used throughout the Cocoa Touch framework, with one example being UIDeviceOrientation. In Swift, there is a protocol, OptionSet, that formalizes this pattern and provides additional convenience. In this recipe, we will explore how to define your own option sets.

Getting ready

In this recipe, we won't be using any components from the previous recipes, so you can create a new playground for this recipe.

How to do it...

Let's rewrite our example from the last recipe, which defined supported device values, to use OptionSet:

```
struct Devices: OptionSet {
    let rawValue: Int

    static let phone = Devices(rawValue: 1 << 0)
    static let tablet = Devices(rawValue: 1 << 1)
    static let watch = Devices(rawValue: 1 << 2)
    static let laptop = Devices(rawValue: 1 << 3)
    static let desktop = Devices(rawValue: 1 << 4)
    static let tv = Devices(rawValue: 1 << 5)
    static let brainImplant = Devices(rawValue: 1 << 6)

    static let none: Devices = []
    static let all: Devices = [.phone, .tablet, .watch, .laptop,
.desktop, .tv, .brainImplant]
    static let stationary: Devices = [.desktop, .tv]
    static let supportsUIKit: Devices = [.phone, .tablet, .tv]
}

let supportedDevices: Devices = [.phone, .tablet, .watch, .tv]]
```

How it works...

The `OptionSet` protocol requires a `rawValue` property, and the convention is to define static constants for each of the options. Additionally, convenient combinations of options can also be defined as static constants, and `OptionSet` provides a convenience initializer, which allows an array of options to be provided, then the options are combined through bitwise addition and stored as one value.

The `OptionSet` protocol provides set-like manipulation and comparison methods that perform the same bitwise operations that we covered in the last recipe:

```
// Contains / AND and comparison
let phoneIsSupported = supportedDevices.contains(.phone)

// Union / OR
let stationaryOrUIKitDevices = Devices.supportsUIKit.union(Devices.
stationary)

// Intersection / AND
let stationaryAndUIKitDevices = Devices.supportsUIKit.
intersection(Devices.stationary)
```

Many of the set methods that we examined in *Chapter 2, Mastering the Building Blocks*, are also provided.

See also

Further information about the `OptionSet` protocol can be found in Apple's documentation on the Swift language at `http://swiftbook.link/docs/optionset`.

Creating custom operators

In an earlier recipe, we looked at some of the advanced operators that Swift offers on top of the common mathematical operators. In this recipe, we will look at how we can create our own operators, enabling very concisely expressive behaviors that feel like part of the language.

The custom operator we will create will be used to append the information in one value to the information in another value, producing a new value that contains the second value, followed by the first. The functionality we are looking to achieve is similar to the `>>` Unix command.

Getting ready

Let's understand how the `>>` Unix command works, and in this recipe, we will implement something similar in Swift using a custom operator.

Since macOS is Unix-based, we can provide Unix commands within *Terminal*. Open up the Terminal application on your Mac:

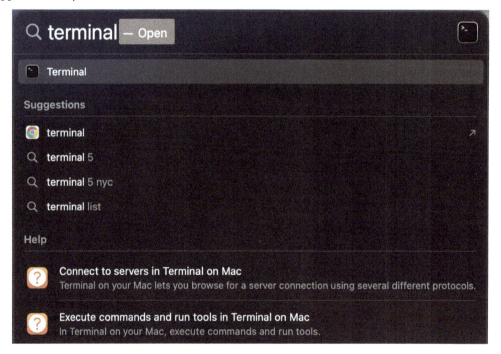

Figure 4.1 – Spotlight search

Type `cd ~/Desktop` and press *Enter* to move to the folder containing all the files and folders on your desktop. Type `touch Tasks.txt` and then press *Enter* to create a blank text file on your desktop called `Tasks.txt`.

To add tasks to our tasks text file we can type the following command, followed by *Enter*:

```
echo "buy milk" >> Tasks.txt
```

If you open the text file on your desktop, you'll see that we have added `buy milk` on the first line.

Enter another task in the same way:

```
echo "mow the lawn" >> Tasks.txt
```

Reopen the `Tasks.txt` file, and you will see that `mow the lawn` has been added on the second line:

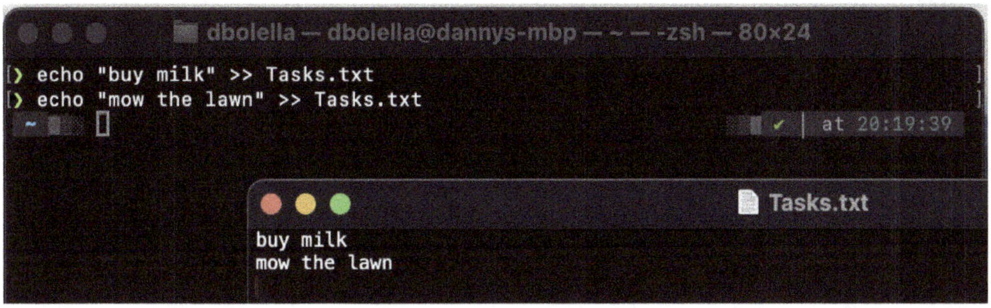

Figure 4.2 – Task result

Add a few more tasks in the same way, and you'll see that each task is appended to the text file on the next line.

The command we issued in Terminal takes the following form:

```
<#What to append#> >> <#Where to append it#>
```

Let's create a similar behavior in Swift; however, we can't use the same command string, `>>`, as this is already defined as bit shifting to the right, so let's make it `>>>`.

Let's create a new playground for this recipe.

How to do it...

We will define and then use a new *append* operator, `>>>`:

1. Declare an infix operator:

    ```
    infix operator >>>
    ```

2. Define the behavior of our operator when used with two strings:

```
func >>> (lhs: String, rhs: String) -> String {
    var combined = rhs
    combined.append(lhs)
    return combined
}
```

3. Define the behavior of our operator when appending a String to an array of strings:

```
func >>> (lhs: String, rhs: [String]) -> [String] {
    var combined = rhs
    combined.append(lhs)
    return combined
}
```

4. Define the behavior of our operator when appending an array of strings to another array of strings:

```
func >>> (lhs: [String], rhs: [String]) -> [String] {
    var combined = rhs
    combined.append(contentsOf: lhs)
    return combined
}
```

5. With these implementations in place, use our new operator to append things:

```
let appendedString = "Two" >>> "One"
print(appendedString) // OneTwo
let appendedStringToArray = "three" >>> ["one", "two"]
print(appendedStringToArray) // ["one", "two", "three"]
let appendedArray = ["three", "four"] >>> ["one", "two"]
print(appendedArray)     // ["one", "two", "three", "four"]
```

Refactor the preceding two operator implementations to use a generic element type for arrays:

```
func >>> <Element>(lhs: Element, rhs: Array<Element>) ->
Array<Element> {
    var combined = rhs
    combined.append(lhs)
    return combined
}

func >>> <Element>(lhs: Array<Element>, rhs: Array<Element>) ->
Array<Element> {
    var combined = rhs
    combined.append(contentsOf: lhs)
    return combined
}
```

6. Use the operator with arrays of any type:

```
let appendedIntToArray = 3 >>> [1, 2]
print(appendedIntToArray) // [1, 2, 3]

let appendedIntArray = [3, 4] >>> [1, 2]
print(appendedIntArray) // [1, 2, 3, 4]
```

We can implement our custom append operator for our own custom types too.

7. Create a `Task` struct and a `TaskList` class to hold it:

```
struct Task {
    let name: String
}

class TaskList: CustomStringConvertible {
    private var tasks: [Task] = []
    func append(task: Task) {
        tasks.append(task)
    }
    var description: String {
        return tasks.map{ $0.name }
                    .joined(separator: "\n")
    }
}
```

8. Extend `TaskList` to add support for our new append operator:

```
extension TaskList {
    static func >>> (lhs: Task, rhs: TaskList) {
        rhs.append(task: lhs)
    }
}
```

9. Append a Task to a TaskList using our custom operator:

```
let shoppingList = TaskList()
Task(name: "get milk") >>> shoppingList print(shoppingList)
Task(name: "get teabags") >>> shoppingList print(shoppingList)
```

How it works...

First, we declared an `infix` operator:

```
infix operator >>>
```

Operators can come in three types:

- **prefix**: Operates on one value and is placed before the value. An example is the NOT operator:

  ```
  let trueValue = !falseValue
  ```

- **postfix**: Operates on one value and is placed after the value. An example is the force unwrap operator:

  ```
  let unwrapped = optional!
  ```

- **infix**: Operates on two values and is placed between them. An example is the addition operator:

  ```
  let five = 2 + 3
  ```

Once we have defined the operator, we can write top-level functions that implement the behavior for each pair of types: one on the left-hand side (LHS) and one on the right-hand side (RHS). Method parameter overloading allows us to specify the operator implementation for multiple-type pairings.

We can define how to append one string to another when our operator is used with strings:

```
func >>> (lhs: String, rhs: String) -> String {
    var combined = rhs
    combined.append(lhs)
    return combined
}
```

We can implement appending a string to an array of strings:

```
func >>> (lhs: String, rhs: [String]) -> [String] {
    var combined = rhs
    combined.append(lhs)
    return combined
}
```

We can also implement appending the elements in an array of strings to another array of strings:

```
func >>> (lhs: [String], rhs: [String]) -> [String] {
    var combined = rhs
    combined.append(contentsOf: lhs)
    return combined
}
```

This allows us to use the operator with strings and arrays of strings, as those are the implementations we defined:

```
let appendedString = "Two" >>> "One" print(appendedString) // OneTwo
let appendedStringToArray = "three" >>> ["one", "two"]
```

```
print(appendedStringToArray) // ["one", "two", "three"]
let appendedArray = ["three", "four"] >>> ["one", "two"]
print(appendedArray)       // ["one", "two", "three", "four"]
```

We can implement our appending operator on every type of array that we think might be useful, or instead, we can implement it as a generic function and have it work for all arrays.

So, we can refactor the preceding two array implementations to use a generic element type:

```
func >>> <Element>(lhs: Element, rhs: Array<Element>) ->
Array<Element> {
    var combined = rhs
    combined.append(lhs)
    return combined
}

func >>> <Element>(lhs: Array<Element>, rhs: Array<Element>) ->
Array<Element> {
    var combined = rhs
    combined.append(contentsOf: lhs)
    return combined
}
```

This allows us to use arrays of integers without having to explicitly define them for integer arrays:

```
let appendedIntToArray = 3 >>> [1, 2]
print(appendedIntToArray) // [1, 2, 3]

let appendedIntArray = [3, 4] >>> [1, 2]
print(appendedIntArray)       // [1, 2, 3, 4]
```

We can also implement it for our own custom types. Let's create `Task` and `TaskList`, which might benefit from using the operator:

```
struct Task {
    let name: String
}

class TaskList: CustomStringConvertible {
    private var tasks: [Task] = []
    func append(task: Task) {
        tasks.append(task)
    }
    var description: String {
    return tasks.map { $0.name }
                .joined(separator: "\n")
```

```
        }
    }
```

We've added `CustomStringConvertible` conformance so that we can easily print out the result.

An alternative to implementing the use of an operator as a top-level function is to declare it within the relevant type as a static function. We'll declare it within an extension on our `TaskList` object, but we could just as easily declare it within the main `TaskList` class declaration:

```
extension TaskList {
    static func >>> (lhs: Task, rhs: TaskList) {
        rhs.append(task: lhs)
    }
}
```

Implementing this within a type has a few advantages: the implementation code is right next to the type itself, making it easier to find, and taking advantage of any values or types that might have a private, or otherwise restricted, access control, which will prevent them from being visible to a top-level function.

Now we can use our `>>>` operator to append a Task to a TaskList:

```
let shoppingList = TaskList()
Task(name: "get milk") >>> shoppingList
print(shoppingList)
Task(name: "get teabags") >>> shoppingList print(shoppingList)
```

We have created a custom operator, to allow more concise and expressive code.

There's more...

Operators don't just work individually – they are often used within the same expression as other operators; the mathematical operators are a helpful example of this:

```
let result = 6 + 8 / 2 / 4
```

The order that these operations are performed in will affect the result. To understand the order in which the operations are performed, we can add brackets that will perform the same function:

```
let result = 6 + ((8 / 2) / 4)
```

In Swift, the decision about how to order operations is made using two concepts – precedence and associativity:

- **Precedence**: This defines how important the operation type is. Therefore, the operations with the highest precedence are performed first; for example, multiplication (*) has higher precedence than addition (+) and is therefore always performed first.

- **Associativity**: This defines which side, left or right, a value should associate itself with for evaluation when it has an operation with the same precedence on either side. This has the effect of defining the order that operations of the same precedence should be evaluated in: left to right or right to left.

Let's use this information to understand the operation ordering of the preceding mathematical operation. We have an expression comprising one addition and two division operations. Division operations have higher precedence than addition operations; therefore, the division operations are evaluated first:

```
let result = 6 + (8 / 2 / 4)
```

We now have two division operations that need to be evaluated before the addition operation. Since both are division operators, they have the same precedence, so we have to look at associativity to know in which order to evaluate them. The division operation has an associativity of left, so they should be evaluated from left to right. Therefore, 8 / 2 is evaluated first and 4 / 4 is evaluated next. This gives us the following:

```
let result = 6 + ((8 / 2) / 4)
```

We need to define precedence and associativity for our custom operator, as the compiler does not currently know how it should be ordered within an expression containing multiple operations. Because of this, the following expression will not compile:

```
let multiOperationArray = [5,6] >>> [3,4] >>> [1,2] + [9,10] >>> [7,8]
print(multiOperationArray)
```

Precedence and associativity are defined within a precedence group, and an operator can either conform to an existing group or one that has been newly defined.

Let's define a new precedence group for our appending operator:

```
precedencegroup AppendingPrecedence {
    associativity: left
    higherThan: AdditionPrecedence
    lowerThan: MultiplicationPrecedence
}
```

Here, we give it the name `AppendingPrecedence` and define its values within curly brackets. We'll set its associativity to the left to match mathematical operations, and to establish precedence, we define that this precedence group is higher than another precedence group and lower than some other precedence groups. For the appending operator, we'll set the precedence to be higher than addition, so it will be evaluated before the addition operators but after the multiplication operators. Both the `AdditionPrecendence` and `MultiplicationPrecedence` groups are defined by the standard library.

Now that we have a precedence group defined, we can ensure that our custom operator conforms to it:

```
infix operator >>> : AppendingPrecedence
```

With precedence and associativity declared, the composite expression previously created will now compile:

```
let multiOperationArray = [5,6] >>> [3,4] >>> [1,2] + [9,10] >>> [7,8]
print(multiOperationArray) // [1,2,3,4,5,6,7,8,9,10]
```

We have defined how our custom operator works alongside other operators, allowing for complex combinations.

See also

Further information about custom operators can be found in Apple's documentation on the Swift language at http://swiftbook.link/docs/custom-operators.

Nesting types and namespacing

In Objective-C, all objects are at the top level and are given a global scope. They can be said to be in the same namespace. This is one reason for the convention among Objective-C developers, including Apple, of prefixing their class names with two- or three-letter identifiers.

These prefix characters allow similarly named classes from different frameworks to be differentiated, for example, UIView from UIKit and SKView from SpriteKit. Swift solves this problem by allowing types to be nested within other types, providing namespacing with nested types and modules.

Any type can be defined as being nested within another type. This allows us to tightly associate one type with another, in addition to providing namespacing, which helps differentiate types with the same name. In this recipe, we will create some nested types to see if it affects how they are referenced.

Getting ready

In this recipe, we won't be using any components from the previous recipes, so you can create a new playground for this recipe.

How to do it...

Let's build a system to monitor a physical device and the user interface that it displays. Both the device and the user interface have the concept of orientation, although each has a differing definition:

1. Define a class to represent the device:

    ```
    class Device {
        enum Category {
    ```

```
            case watch
            case phone
            case tablet
        }
        enum Orientation {
            case portrait
            case portraitUpsideDown
            case landscapeLeft
            case landscapeRight
        }
        let category: Category
        var currentOrientation: Orientation = .portrait
        init(category: Category) {
            self.category = category
        }
    }
```

Within this class, we have defined two enums, which only have value when used in relation to the Device class. Nesting the types also allows us to simplify the names of these types. It would be customary to name them DeviceCategory and DeviceOrientation to avoid confusion, but since they are nested, we can remove the Device prefix.

Any use of the nested types within the type that contains them can be done without any qualifiers; however, this is not the case for use outside of the containing type.

2. Access nested types from outside the containing type using dot syntax:

```
let phone = Device(category: .phone)
let desiredOrientation: Device.Orientation = .portrait
let phoneHasDesiredOrientation = phone.currentOrientation ==
desiredOrientation
```

To reference a nested type, we must first specify the containing type, so the Orientation enum, within the Device class, becomes Device.Orientation.

3. Define a struct to represent a user interface:

```
struct UserInterface {
    struct Version {
        let major: Int
        let minor: Int
        let patch: Int
    }
    enum Orientation {
        case portrait
        case landscape
    }
```

```
    let version: Version
    var orientation: Orientation
}
```

Our `UserInterface` struct also includes a nested `Orientation` enum, but as these two enums lie in different namespaces, there is no naming conflict. As before, the nested types can be used without any qualifiers in the containing type.

4. Let's see how these two nested types can be used in conjunction with one another. Create a function to convert from device orientation to user interface orientation:

```
func uiOrientation(for deviceOrientation: Device.Orientation) ->
UserInterface.Orientation {
    switch deviceOrientation {
    case Device.Orientation.portrait, Device.Orientation.
portraitUpsideDown:
        return UserInterface.Orientation.portrait
    case Device.Orientation.landscapeLeft, Device.Orientation.
landscapeRight:
        return UserInterface.Orientation.landscape
    }
}
let phoneUIOrientation = uiOrientation(for: phone.
currentOrientation)
print(phoneUIOrientation) // UserInterface.Orientation.portrait
```

How it works...

Our orientation conversion function specifies the full `enum` case for the `switch` statement and the return statements, for example:

```
Device.Orientation.portrait UserInterface.Orientation.portrait
```

However, as we've seen previously, when the compiler knows the type of the enum, only the case needs to be specified; the enum type can be removed. For our function, the input parameter type is `Device.Orientation` and the return type is `UserInterface.Orientation`, so the compiler does know enum types, and therefore we can remove the types:

```
func uiOrientation(for deviceOrientation: Device.Orientation) ->
UserInterface.Orientation {
    switch deviceOrientation {
    case .portrait, .portraitUpsideDown:
        return .portrait
    case .landscapeLeft, .landscapeRight:
        return .landscape
    }
}
```

Note that the `switch` case contains `.portrait` and returns `.portrait`, but these are cases from different enums, and the compiler knows the difference.

There's more...

We've seen how namespacing separates types nested within different containing types, but types within modules are also namespaced. This allows you to name your types without fear of collision with types in other modules.

Let's imagine that we are building an app for hospitals to keep track of their events and resources. As part of this, we create a class to represent surgical operations that we intend to track:

```
class Operation {
    let doctorsName: String
    let patientsName: String
    init(doctorsName: String, patientsName: String) {
        self.doctorsName = doctorsName
        self.patientsName = patientsName
    }
}
```

There is another class called `Operation`, provided by the `Foundation` framework, that can be used to execute and manage a long-running task. We can use both types of `Operation` side by side because the `Foundation` framework is exposed as a module, and so the long-running `Operation` class can be used by referencing the `Foundation` module:

```
import Foundation
let medicalOperation = Operation(doctorsName: "Dr. Crusher",
patientsName: "Commander Riker")
let longRunningOperation = Foundation.Operation()
```

We've seen how you can disambiguate two types with the same name, using the module they are within.

See also

Further information about nested types can be found in Apple's documentation on the Swift language at `http://swiftbook.link/docs/nested-types`.

5

Beyond the Standard Library

Apple's intention when open sourcing Swift was to provide a cross-platform, general-purpose programming language that is ready to use. The Swift standard library provides core language features and common collection types. However, the library does not provide everything needed to get up and running.

Therefore, Apple provides a framework called **Foundation** to help you perform common programming tasks that aren't covered by the core Swift language and the standard library.

The Foundation framework that you will use when developing for Apple platforms is **closed sourced**, which means the underlying code is not accessible and only the API is visible. However, when Apple open sourced Swift and made it available for Linux, it became necessary to provide the Foundation framework as well. To this end, Apple has released an open source, Swift-based version of Foundation as a core library, available here: `https://github.com/apple/swift-corelibs-foundation`.

By the end of this chapter, you will have gained experience using some of the most common tools in Foundation, particularly fetching and massaging data from the network/web!

In this chapter, we will cover the following recipes:

- Comparing dates with Foundation
- Fetching data with `URLSession`
- Working with JSON
- Working with XML

Technical requirements

All the code for this chapter can be found in this book's GitHub repository at `https://github.com/PacktPublishing/Swift-Cookbook-Third-Edition/tree/main/Chapter%205`.

Comparing dates with Foundation

This recipe will focus on one area of Foundation that is very widely used, that is, date and time manipulation and formatting.

We will create a function that determines how long there is until Halloween and return this information as a string that can be displayed to a user.

Getting ready

Create a new iOS playground and import the Foundation framework at the top of the playground:

```
import Foundation
```

How to do it...

Let's create a function that will return a string telling us how long there is until Halloween. We can then print the result:

1. Define the function:

   ```
   func howLongUntilHalloween() -> String {

   }
   ```

2. Within the function, get the current calendar and time zone:

   ```
   let calendar = Calendar.current
   let timeZone = TimeZone.current
   ```

3. Get the current date and time and use the calendar to get the current year:

   ```
   let now = Date()
   let yearOfNextHalloween = calendar.component(.year, from: now)
   ```

4. Define date components that correspond to midnight on Halloween:

   ```
   var components = DateComponents(
       calendar: calendar,
       timeZone: timeZone,
       year: yearOfNextHalloween,
       month: 10,
       day: 31,
       hour: 0,
       minute: 0,
       second: 0)
   ```

5. Get a `Date` object from those components:

```
var halloween = components.date!
```

6. If we have already passed Halloween for this year, we need to adjust the component to refer to Halloween of the next year:

```
// If we have already had Halloween this year, then we need to
use Halloween next year.
if halloween < now {
    components.year = yearOfNextHalloween + 1
    halloween = components.date!
}
```

7. Create `DateComponentsFormatter` to format how the time until Halloween is displayed:

```
let componentFormatter = DateComponentsFormatter()
componentFormatter.unitsStyle = .full
componentFormatter.allowedUnits = [.month, .day,
.hour, .minute, .second]
```

8. Use `DateComponentsFormatter` to return a string for the time between now and next Halloween:

```
return componentFormatter.string(from: now, to: halloween)!
```

9. Below the `howLongUntilHalloween` function, use the following function to create a string, and print the outcome:

```
let timeUntilHalloween = howLongUntilHalloween()
print("Time until Halloween: \(timeUntilHalloween)")
```

How it works...

In *step 1*, we created our `howLongUntilHalloween` function, then in *step 2*, we fetched the currently set calendar and time zone as these will be needed for the date calculations to come:

```
let calendar = Calendar.current
let timeZone = TimeZone.current
```

While retrieving the current time zone is self-explanatory, it is not immediately obvious what the `Calendar` type represents and why we need to retrieve it.

How dates are represented is not as universally agreed on as you might believe. Certain time components are mostly universal, such as the length of years and days, as they are connected to astronomical events, such as the time it takes for the Earth to perform one revolution of the Sun, and for the Earth to complete one rotation around its own axis, respectively. However, other time components, such as months and weeks and how years are numbered, are rooted in the culture that created them.

> **Interesting fact**
>
> The calendar used throughout Europe and most of the world is known as the Gregorian calendar, introduced in 1582 by Pope Gregory XIII, replacing the Julian calendar. There are about 40 different calendars currently in use around the world, including Gregorian, Chinese, Hebrew, Islamic, Persian, Ethiopian, and Balinese Pawukon.

The way in which we present how long there is until Halloween will depend on the calendar that is relevant to the user. This is why we ask for the current calendar, which the user can change if they want a different representation.

Our next task is to get the current date and time:

```
let now = Date()
```

In *step 3*, the default initializer for the Date value type uses the current date and time as its value. Note that this date value is set at the point of creation; it does not continually update with the current date and time.

In this step, we get a date and time for the next Halloween. We know the time, day, and month of Halloween, so to construct a date for Halloween, we just need to know the year. There is a method on Calendar called component that allows us to retrieve specific components from a Date value:

```
let yearOfNextHalloween = calendar.component(.year, from: now)
```

We now have the current year, within the user's current calendar; we can use it to create the Halloween date.

In *step 4*, we create an instance of DateComponents, passing in the calendar, time zone, and the fact that we are defining October 31 at midnight, for the current year:

```
var components = DateComponents(
    calendar: calendar,
    timeZone: timeZone,
    year: yearOfNextHalloween,
    month: 10,
    day: 31,
    hour: 0,
    minute: 0,
    second: 0)
```

In *step 5*, we create a Date object from DateComponents. This is of an optional type as we may not have provided enough information to the components to generate a date; however, since we know that we have the information in this instance, we can force-unwrap this optional:

```
var halloween = components.date!
```

Next, we need to handle an edge case; what if we have already had Halloween this year? For example, let's imagine that the current date is November 2, 2022. We are trying to find the date of the next Halloween, but if we use the current year, we will get October 31, 2022, which is the Halloween just gone. So, in *step 6*, we add *1* to the current year to get next Halloween, October 31, 2023:

```
if halloween < now {
    components.year = yearOfNextHalloween + 1
    halloween = components.date!
}
```

To account for this, we check whether the Halloween for this year is before now; if it is, we bump the year component to next year and recreate the Halloween date from `DateComponent`.

We now have the current date and the next Halloween date, and Foundation provides functionality to calculate the time difference between two dates and format it to display to a user, through the use of `DateComponentsFormatter`.

In *step 7*, we create `DateComponentsFormatter`, and set `unitStyle` to `full`, which will provide a string using the full unit name, without abbreviation. We configure how we want the date and time divided for display, using `allowedUnits`:

```
let componentFormatter = DateComponentsFormatter()
componentFormatter.unitsStyle = .full
componentFormatter.allowedUnits = [.month, .day, .hour,
.minute, .second]
```

In *step 8*, we can retrieve a string from the formatter that describes the time between the two dates given, with the settings provided to the formatter. Since `DateComponentsFormatter` returns an optional string, we unwrap and return it:

```
return componentFormatter.string(from: now, to: halloween)!
```

Our `howLongUntilHalloween` method will provide a string describing how long until Halloween, which we can then print out.

See also

There is a lot more to discover in Foundation, so check out the documentation for further functionality:

- Swift documentation for Foundation: `https://developer.apple.com/documentation/foundation`
- Open source repository for Foundation: `https://github.com/apple/swift-corelibs-foundation`

Fetching data with URLSession

Every app worth building will need to send or receive information from the internet at some point, and therefore, networking support is a critical part of any development platform. In Swift, this support for networking is provided by the Foundation framework.

When we need to retrieve information from the internet, we send out a request to a server on the internet, and that server sends a response that hopefully contains the information we requested.

In this recipe, we will learn how to send network requests and receive a response using the Foundation framework.

Getting ready

It is helpful to know about the different components that Foundation provides that deal with networking and what they do:

- URL: The address of a resource on a remote server. It contains information about the server and where the resource can be found on the server.

- URLRequest: Represents the request that will be made to the remote server. Defines the URL of the resource, how the request should be sent, metadata in the form of headers, and data that should be sent with it.

- URLSession: Manages the communication with remote servers, holds the configuration for that communication, and creates and optimizes the underlying connections.

- URLSessionDataTask: An object that manages the state of the request and delivers the response.

- URLResponse: Holds the metadata of the response from the remote server.

How to do it...

Let's use these networking tools to retrieve an image from a remote server:

Import PlaygroundSupport and set up indefinite execution for this playground:

```
import PlaygroundSupport
PlaygroundPage.current.needsIndefiniteExecution = true
```

1. Import Foundation and create an instance of URLSession:

```
import Foundation
let config = URLSessionConfiguration.default
let session = URLSession(configuration: config)
```

2. Next, we will construct a request for a remote image:

```
let urlString = "https://imgs.xkcd.com/comics/api.png"
let url = URL(string: urlString)!
let request = URLRequest(url: url)
```

3. Now that we have our `URLRequest`, we can create a data task to retrieve the image from the remote server:

```
let task = session.dataTask(with: request, completionHandler: {
(data, response, error) in
    // More code to follow
})
```

4. We will take the image data and put it in a `UIImage` object to display it. We need to import the `UIKit` framework, which provides `UIImage`. So, let's import `UIKit` at the top of the playground:

```
import UIKit
```

5. Check for image data in the completion handler and create a `UIImage` object:

```
let task = session.dataTask(with: request) { (data, response,
error) in
    guard let imageData = data else {
        return // No Image, handle error
    }
    _ = UIImage(data: imageData)
}
```

6. Call `resume` on the task to start it:

```
task.resume()
```

How it works...

Let's walk through the previously mentioned steps to understand what we are doing:

```
import PlaygroundSupport
PlaygroundPage.current.needsIndefiniteExecution = true
```

Playgrounds execute the code they contain from top to bottom. When the end of the playground page is reached, the playground stops executing. In our example, the task is created and started, but then the playground reaches the end of the page before the image has been fully retrieved because this happens asynchronously. If the playground were to stop execution here, the completion handler would never be executed, and we wouldn't see the image. This isn't a problem in a normal app that is continually running while it is in use; it is just specific to how Swift playgrounds work.

To solve this, we need to tell the playground that we don't want it to stop executing when it reaches the end of the page, and instead, it should run indefinitely while we wait for the response to be received. This is done by importing the `PlaygroundSupport` framework.

in *step 1*, `needsIndefiniteExecution` is set to `true` on the current `PlaygroundPage`:

```
import Foundation
let config = URLSessionConfiguration.default
let session = URLSession(configuration: config)
```

In *step 2*, when creating `URLSession`, we pass in a `URLSessionConfiguration` object, which allows configuring the time it takes for a request to time out and cache responses, among other things. For our purposes, we will just use the default configuration:

```
let urlString = "https://imgs.xkcd.com/comics/api.png"
let url = URL(string: urlString)!
let request = URLRequest(url: url)
```

In *step 3*, we will be requesting the image from the excellent webcomic *XKCD* (`http://xkcd.com`). We can create the URL from a string, and then create a `URLRequest` request from the URL:

```
let task = session.dataTask(with: request, completionHandler: { (data,
response, error) in
    // More code to follow
})
```

In *step 4*, we do not create a data task directly; instead, we ask our `URLSession` instance to create the data task, and we pass in `URLRequest` and a completion handler. The completion handler will be fired once a response has been received from the remote server or an error has occurred.

The completion handler has three inputs, each with its own data type (shown after the colon), which are all optional to use:

- `data: Data` - The data returned in the body of the response; if our request was successful, this will contain our image data.

- `response: URLResponse` - The response metadata, including response headers. If the request was over HTTP/HTTPS, then this will be `HTTPURLResponse`, which will contain the HTTP status code.

- `error: Error` - If the request was unsuccessful, due to a network issue, for example, this value will have the error, and the data and response value will be nil. If the request was successful, this error value will be nil:

```
let task = session.dataTask(with: request) { (data, response, error) in
    guard let imageData = data else {
        return // No Image, handle error
```

```
    }
    _ = UIImage(data: imageData)
}
```

In *step 6*, we check for response data and turn it into an image. To do this, we will need to construct a UIImage object from the data. UIImage is a class that represents an image on iOS and can be found in the UIKit framework. So, we also needed to import UIKit at the top of the playground, as we did in *step 5*.

> **Note**
>
> Since we don't plan on doing anything with the image in this example, we are just going to view it in a playground preview; the compiler will complain if we assign it to a value that is never used. Therefore, we replace a normal value assignment with _, which allows the UIImage object to be generated without it being assigned to anything.

In *step 7*, we created the data task to retrieve the image, but we need to actually start the task to make the request. To do that, we call resume() on the task.

```
task.resume()
```

When we run the playground, you will eventually see that the image value has been populated in the playground's right sidebar, and you can click on the preview icon to see the image that has been downloaded:

Figure 5.1 – Retrieved image displayed in the playground timeline

See also

- Further information about networking can be found in Apple's networking overview: `http://swiftbook.link/docs/networking`

- More information can also be found in Apple's URLSession programming guide: `http://swiftbook.link/docs/urlsession-guide`

Working with JSON

As discussed in the last recipe, almost every app will need to exchange information with the internet at some point, and in that recipe, we retrieved an image from a remote server. Very often, your app will need to retrieve more varied data, perhaps relating to the result of a search, or information about a shared state held on the server.

This information can be represented in any number of ways, but one of the most common ways is as JavaScript Object Notation (JSON), which is a text-based structure for representing information. A JSON object contains key-value pairs, where the keys are strings and the values can be strings, numbers, Booleans, null, other objects, or arrays.

For example, information about a person could be expressed with this JSON object:

```
{
    "name": {
        "givenName": "Keith",
        "middleName": "David",
        "familyName": "Moon"
    },
    "age": 40,
    "heightInMetres": 1.778,
    "isBritish": true,
    "favouriteFootballTeam": null
}
```

The following is an example of an array of JSON objects:

```
[
    {
        "name": {
            "givenName": "Keith",
            "middleName": "David",
            "familyName": "Moon"
        },
        "age": 40,
        "heightInMetres": 1.778,
```

```
        "isBritish": true,
        "favouriteFootballTeam": null
    },
    {

        "name": {
            "givenName": "Alissa",
            "middleName": "May",
            "familyName": "Moon"
        },
        "age": 35,
        "heightInMetres": 1.765,
        "isBritish": false,
        "favouriteFootballTeam": null

    }
]
```

Foundation provides tools for reading information from and writing information as JSON data. In this recipe, we will interact with a JSON-based **Application Programming Interface (API)**, to both send and receive information.

Getting ready

Our goal is to interact with the GitHub API and create an issue for this book's repository. A full explanation of Git and GitHub is beyond the scope of this book; suffice it to say that it's a service that stores versioned copies of your source code. Resources relevant to this book are stored in repositories on GitHub, and a GitHub user can create issues that serve as bug reports or feature requests.

If you don't already have one, then you will need to sign up for a GitHub account:

1. Go to https://github.com.
2. Click **Sign up** and enter your email address to continue.
3. Once you have created a GitHub account, you will need to create a personal access token, which we will use to authenticate some of the requests to the GitHub API.
4. To create a personal access token, use the following steps:
5. Go to the settings page (https://github.com/settings/tokens) and click on **Generate new token**.
6. Give the token a name and check the box next to **repo**:

New personal access token

Personal access tokens function like ordinary OAuth access tokens. They can be used instead of a password for Git over HTTPS, or can be used to authenticate to the API over Basic Authentication.

Note

Swift Cookbook

What's this token for?

Select scopes

Scopes define the access for personal tokens. Read more about OAuth scopes.

☑ **repo**	Full control of private repositories
☑ repo:status	Access commit status
☑ repo_deployment	Access deployment status
☑ public_repo	Access public repositories
☑ repo:invite	Access repository invitations
☑ security_events	Read and write security events

Figure 5.2 – Creating a personal access token

7. Click on **Generate token** at the bottom of the page. You will now see your newly generated personal access token.

8. Copy this token and paste it somewhere, as we will need it later:

Personal access tokens

Generate new token | Revoke all

Tokens you have generated that can be used to access the GitHub API.

Make sure to copy your new personal access token now. You won't be able to see it again!

✓ ▓▓▓▓▓▓▓▓▓▓▓▓▓▓▓▓▓▓▓▓▓▓▓▓▓ 📋   Delete

Figure 5.3 – The generated personal access token

How to do it...

To create our issue, we will first retrieve all of Packt Publishing's public repositories, and then find the relevant repository for this book. We will then create a new issue in this repository.

As in the preceding recipe, we will need a URLSession object to perform our requests, and we need to tell the playground not to finish executing when it reaches the end of the playground, so that we have a chance to handle our request:

```
import Foundation
import PlaygroundSupport
PlaygroundPage.current.needsIndefiniteExecution = true
let config = URLSessionConfiguration.default
let session = URLSession(configuration: config)
```

Our first step is to fetch all the public repositories for a given user:

1. Let's create a function to do that:

```
func fetchRepos(forUsername username: String) {
    let urlString = "https://api.github.com/users/\(username)/
repos"
    let url = URL(string: urlString)!
    var request = URLRequest(url: url)
    request.setValue("application/
vnd.github.v3+json", forHTTPHeaderField: "Accept")
    let task = session.dataTask(with: request) { (data,
response, error) in
        // More code to follow
    }
    task.resume()
}
```

You will note that after creating URLRequest, we set an HTTP header; this particular header ensures that we will always get back version 3 of the GitHub API.

> **Note**
>
> We know from the GitHub API documentation (https://developer.github.com/v3/) that this response data is in JSON format. We need to parse the JSON data to turn it into something that we can use; enter JSONSerialization. JSONSerialization is part of the Foundation framework and provides class methods for turning Swift dictionaries and arrays into JSON data (known as serialization) and back again (known as deserialization).

2. Let's use JSONSerialization to turn our JSON response data into something more useful:

```
func fetchRepos(forUsername username: String) {
    let urlString = "https://api.github.com/users/\(username)/
repos"
    let url = URL(string: urlString)!
    var request = URLRequest(url: url)
    request.setValue("application/vnd.github.v3+json",
```

```
forHTTPHeaderField: "Accept")
    let task = session.dataTask(with: request) { (data,
response, error) in
        // Once we have handled this response, the Playground can
finish executing
        defer {
            PlaygroundPage.current.finishExecution()
        }
        // First unwrap the optional data
        guard let jsonData = data else {
            // If the data is nil, there was probably a network
error
            print(error ?? "Network Error")
            return
        }
        do {
            // Deserialisation can throw an error, so we have to
use `try` and catch the errors
            let deserialised = try JSONSerialization.
jsonObject(with: jsonData, options: [])
            print(deserialised)
        } catch {
            print(error)
        }
    }
    task.resume()
}
```

3. Now, let's fetch the public Packt repositories by executing our function and passing PacktPublishing in as the GitHub username:

```
fetchRepos(forUsername: "PacktPublishing")
```

Once executed, the print output should look like this:

```
(
{
"archive_url" = "https://api.github.com/repos/PacktPublishing/-.NET-Core-Microservices/{archive_format}{/ref}";
archived = 0;
"assignees_url" = "https://api.github.com/repos/PacktPublishing/-.NET-Core-Microservices/assignees{/user}";
"blobs_url" = "https://api.github.com/repos/PacktPublishing/-.NET-Core-Microservices/git/blobs{/sha}";
"branches_url" = "https://api.github.com/repos/PacktPublishing/-.NET-Core-Microservices/branches{/branch}";
"clone_url" = "https://github.com/PacktPublishing/-.NET-Core-Microservices.git";
"collaborators_url" = "https://api.github.com/repos/PacktPublishing/-.NET-Core-Microservices/collaborators{/collaborator}";
"comments_url" = "https://api.github.com/repos/PacktPublishing/-.NET-Core-Microservices/comments{/number}";
"commits_url" = "https://api.github.com/repos/PacktPublishing/-.NET-Core-Microservices/commits{/sha}";
"compare_url" = "https://api.github.com/repos/PacktPublishing/-.NET-Core-Microservices/compare/{base}...{head}";
"contents_url" = "https://api.github.com/repos/PacktPublishing/-.NET-Core-Microservices/contents/{+path}";
"contributors_url" = "https://api.github.com/repos/PacktPublishing/-.NET-Core-Microservices/contributors";
"created_at" = "2019-05-03T10:54:14Z";
"default_branch" = master;
"deployments_url" = "https://api.github.com/repos/PacktPublishing/-.NET-Core-Microservices/deployments";
description = " .NET Core Microservices, published by [Packt]";
disabled = 0;
"downloads_url" = "https://api.github.com/repos/PacktPublishing/-.NET-Core-Microservices/downloads";
```

Figure 5.4 – Public GitHub repositories API response

JSONSerialization has turned our JSON data into familiar arrays and dictionaries that can be used to retrieve the information we need in the normal way. The JSON data is deserialized with the Any type, as the JSON can have a JSON object or an array at its root.

Since, from the preceding output, we know that the response has an array of JSON objects at its root, we need to turn the value from type Any to an array of dictionaries of the [String: Any] type. This is referred to as casting from one type to another, which we can do by using the as keyword and then specifying the new type. This keyword can be used in three different ways:

- as will perform a trivial cast. This is possible if the existing type is synonymous with the intended type, for instance, casting from a subclass to a superclass.

- as? will conditionally perform a cast, returning an optional value. If it is not possible to represent the value as the intended type, the value will be nil.

- as! will perform a forced cast. If it is not possible to represent the value as the intended type, you will get a crash.

So, let's cast the deserialized data to an array of dictionaries with string keys, with the [[String: Any]] type:

```
func fetchRepos(forUsername username: String) {
    //...
    let task = session.dataTask(with: request) { (data, response,
error) in
        //...
        do {
            // Deserialisation can throw an error, so we have to `try`
and catch the errors
            let deserialised = try JSONSerialization.jsonObject(with:
jsonData, options: [])
            print(deserialised)
            // As `deserialised` has type `Any` we need to cast
            guard let repos = deserialised as? [[String: Any]] else {
                print("Unexpected Response")
                return
            }
            print(repos)
        } catch {
            print(error)
        }
    }
    //...
}
```

Now, we have an array of dictionaries for the repositories in the API response, which we need to provide as input for this function. A common pattern for providing results for asynchronous work is to provide a completion handler as a parameter. A completion handler is a closure that can be executed once the asynchronous work is completed.

Since the output we want to provide is the array of repository dictionaries, we will define this as an input for the closure if the request was successful, and an error if it wasn't:

```swift
func fetchRepos(forUsername username: String, completionHandler:
@escaping ([[String: Any]]?, Error?) -> Void) {
    let urlString = "https://api.github.com/users/\(username)/repos"
    let url = URL(string: urlString)!
    var request = URLRequest(url: url)
    request.setValue("application/vnd.github.v3+json",
forHTTPHeaderField: "Accept")
    let task = session.dataTask(with: request) { (data, response,
error) in
        // Once we have handled this response, the Playground can
finish executing
        defer {
            PlaygroundPage.current.finishExecution()
        }
        // First unwrap the optional data
        guard let jsonData = data else {
            // If the data is nil, there was probably a network
error
            completionHandler(nil, ResponseError.
requestUnsuccessful)
            return
        }

        do {
        // Deserialisation can throw an error, so we have to `try`
and catch the errors
            let deserialised = try JSONSerialization.
jsonObject(with: jsonData, options: [])
            // As `deserialised` has type `Any` we need to cast
            guard let repos = deserialised as? [[String: Any]] else
{
                completionHandler(nil, ResponseError.
unexpectedResponseStructure)
                return
            }
            completionHandler(repos, nil)
        } catch {
            completionHandler(nil, error)
```

```
        }
    }
    task.resume()
}
```

Now, whenever an error is generated, we execute `completionHandler`, passing in the error and `nil` for the result. Also, when we have the repository results, we execute the completion handler, passing in the parsed JSON and `nil` for the error.

We passed in a few new errors in the preceding code, so let's define those errors:

```
enum ResponseError: Error {
    case requestUnsuccessful
    case unexpectedResponseStructure
}
```

This changes how we call this `fetchRepos` function:

```
fetchRepos(forUsername: "PacktPublishing") { (repos, error) in
    if let repos = repos {
        print(repos)
    } else if let error = error {
        print(error)
    }
}
```

Now that we have retrieved the details of the public repositories, we will submit an issue to the repository for this chapter. This issue can be any feedback you would like to give about this book; it can be a review, a suggestion for new content, or you can tell me about a Swift project you are currently working on.

This request to the GitHub API will be authenticated against your user account and, therefore, we will need to include details of the personal access token that we created at the beginning of this recipe. There are a number of ways to authenticate requests to the GitHub API, but the simplest is basic authentication, which involves adding an authorization string to the request header.

Let's create a method to format the personal access token correctly for authentication:

```
func authHeaderValue(for token: String) -> String {
    let authorisationValue = Data("\(token):x-oauth-basic".utf8).
base64EncodedString()
    return "Basic \(authorisationValue)"
}
```

Next, let's create our function to submit our issue. From the API documentation at `https://developer.github.com/v3/issues/#create-an-issue`, we can see that unless you have push access, you can only create an issue with the following components:

- `title` (required)

- `body` (optional)

So, our function will take this information as input, along with the repository name and username:

```
func createIssue(inRepo repo: String, forUser user: String, title:
String, body: String?) {
    // More code to follow
}
```

Creating an issue is achieved by sending a POST request, and information about the issue is provided as JSON data in the request body. To create our request, we can use `JSONSerialization`, but we will take our intended JSON structure and serialize it into `Data` this time:

```
func createIssue(inRepo repo: String, forUser user: String, title:
String, body: String?) {
    // Create the URL and Request
    let urlString = "https://api.github.com/repos/\(user)/\ (repo)/
issues"
    let url = URL(string: urlString)!
    var request = URLRequest(url: url)
    request.httpMethod = "POST"
    request.setValue("application/vnd.github.v3+json",
forHTTPHeaderField: "Accept")
    let authorisationValue = authHeaderValue(for: <#your personal
access token>)
    request.setValue(authorisationValue, forHTTPHeaderField:
"Authorization")
    // Put the issue information into the JSON structure required
    var json = ["title": title]
    if let body = body {
        json["body"] = body
    }
    // Serialise the json into Data. We can use try! as we know it is
valid JSON.
    // Just be aware that the this will fail if provided value can't
be converted into valid JSON.
    let jsonData = try! JSONSerialization.data(withJSONObject: json,
options: .prettyPrinted)
    request.httpBody = jsonData
    session.dataTask(with: request) { (data, response, error) in
```

```
                // More code to follow
        }
}
```

As with the previous API request, we need a way to provide the result of creating the issue, so let's provide a completion handler, try to deserialize the response, and provide it to the `completion` handler:

```
func createIssue(inRepo repo: String, forUser user: String, title:
String, body: String?, completionHandler: @escaping ([String: Any]?,
Error?) -> Void) {
    //...
    session.dataTask(with: request) { (data, response, error) in
        guard let jsonData = data else {
            completionHandler(nil, ResponseError.
requestUnsuccessful)
            return
        }
        do {
            // Deserialisation can throw an error, so we have to
`try` and catch the errors
            let deserialised = try JSONSerialization.
jsonObject(with: jsonData, options: [])
            // As `deserialised` has type `Any` we need to cast
            guard let createdIssue = deserialised as? [String: Any]
else {
                completionHandler(nil, ResponseError.
unexpectedResponseStructure)
                return
            }
            completionHandler(createdIssue, nil)
        } catch {
            completionHandler(nil, error)
        }
    }
}
```

The API response to a successfully created issue provides a JSON representation of that issue. Our function will return this representation if it was successful, or an error if it was not.

Now that we have a function to create issues in a repository, it's time to use it to create an issue:

```
createIssue(inRepo: "Swift-Cookbook-Third-Edition",
        forUser: "PacktPublishing",
        title: <#The title of your feedback#>,
        body: <#Extra detail#>) { (issue, error) in
    if let issue = issue {
```

```
        print(issue)
    } else if let error = error {
        print(error)
    }
}
```

> **Note**
>
> I will check these created issues, so please provide genuine feedback on this book. How have you found the content? Too detailed? Not detailed enough? Anything I've missed or not fully explained? Any questions that you have? This is your opportunity to let me know!

There's more...

When we created our completion handlers, we gave them two inputs: the successful result (either the repository information or the created issue) or an error if there is a failure. Both these values are optional; one will be nil, and the other has a value. However, this convention is not enforced by the language, and a user of this function will have to consider the possibility that it may not be the case. What should the user of this function do if the fetchRepos function fires the completion handler with non-nil values for both the repository and the error? What if both are nil?

The user of this function, without viewing the function's internal code, can't be sure that this won't happen, which means they may need to write functionality and tests to account for this possibility, even though it may never happen.

It would be better if we could more accurately represent the intended behavior of our function, providing the user with a clear indication of the possible outcomes and leaving no room for ambiguity. We know that there are two possible outcomes from calling the function: it will either succeed and return the relevant value, or it will fail and return an error to indicate the reason for the failure.

Instead of optional values, we can use an enum to represent these possibilities, and the Foundation framework provides a generic enum for this purpose, called Result.

The Result enum has a success case, which has an associated type for a successful result, and a failure case with an associated type for the relevant error. Both associated types are defined as generic constraints, with the failure type needing to conform to the Error protocol.

We can now define the success and failure states and use associated values to hold the value that is relevant for each state, which is the repository information for the success state and the error for the failure state.

Now, let's amend our fetchRepos function to provide a Result enum in completionHandler:

```
func fetchRepos(forUsername username: String, completionHandler: @
escaping (Result<[[String: Any]], Error?) -> Void) {
    //...
    let task = session.dataTask(with: request) { (data, response,
```

```
error) in
            //...
            // First unwrap the optional data
            guard let jsonData = data else {
            // If the date is nil, there was probably a network error
                completionHandler(.failure(ResponseError.
requestUnsuccessful))
                return
            }

            do {
                // Deserialisation can throw an error, so we have to
`try` and catch the errors
                let deserialised = try JSONSerialization.
jsonObject(with: jsonData, options: [])
                // As `deserialised` has type `Any` we need to cast
                guard let repos = deserialised as? [[String: Any]] else
{
                    let error = ResponseError.
unexpectedResponseStructure
                    completionHandler(.failure(error))
                    return
                }
                completionHandler(.success(repos))
            } catch {
                completionHandler(.failure(error))
            }
        }
        task.resume()
}
```

We need to update how we call the `fetchRepos` function:

```
fetchRepos(forUsername: "PacktPublishing", completionHandler: { result
in
    switch result {
    case .success(let repos):
        print(repos)
    case .failure(let error):
        print(error)
    }
})
```

We now use a `switch` statement instead of `if/else`, and we get the added benefit that the compiler will ensure that we have covered all possible outcomes.

Having made this improvement to the `fetchRepos` function, we can similarly improve the `createIssue` function:

```
func createIssue(inRepo repo: String, forUser user: String, title:
String, body: String?, completionHandler: @escaping (Result<[[String:
Any]], Error?) -> Void) {
    //...
    let task = session.dataTask(with: request) { (data, response,
error) in
        guard let jsonData = data else {
            let error = ResponseError.requestUnsuccessful
            completionHandler(.failure())
            return
        }


        do {
// Deserialisation can throw an error, so we have to `try` and catch
the errors
            let deserialised = try JSONSerialization.
jsonObject(with: jsonData, options: [])

            // As `deserialised` has type `Any` we need to cast
            guard let createdIssue = deserialised as? [String: Any]
else {
                let error = ResponseError.
unexpectedResponseStructure
                completionHandler(.failure(error))
                return
            }
            completionHandler(.success(createdIssue))
        } catch {
            completionHandler(.failure(error))
        }
    }
    task.resume()
}
```

Lastly, we need to update the contents of the completion handler that we provide to the `createIssue` function:

```
createIssue(inRepo: "Swift-Cookbook-Third-Edition",
        forUser: "PacktPublishing",
        title: <#The title of your feedback#>,
        body: <#Extra detail#>) { result in
    switch result {
    case .success(let issue):
```

```
        print(issue)
    case .failure(let error):
        print(error)
    }
}
```

Working with JSON data and extracting relevant information from it can be frustrating. Consider the JSON response for our `fetchRepos` function:

```
[
    {
        "id": 68144965,
        "name": "JSONNode",
        "full_name": "keefmoon/JSONNode",
        "owner": {
            "login": "keefmoon",
            "id": 271298,
            "avatar_url": "https://avatars.githubusercontent.
com/u/271298?v=3",
            "gravatar_id": "",
            "url": "https://api.github.com/users/keefmoon",
            "html_url": "https://github.com/keefmoon",
            "followers_url": "https://api.github.com/users/
keefmoon/followers",
            //... Some more URLs
            "received_events_url": "https://api.github.com/users/
keefmoon/received_events",
            "type": "User",
            "site_admin": false
        },
        "private": false,
        //... more values
    },
    //... more repositories
]
```

If we want to get the username for the owner of the first repository, we need to deserialize the JSON and then conditionally unwrap multiple nested layers to get the username string:

```
let jsonData = //... returned from the network
guard let deserialised = try? JSONSerialization.jsonObject(with:
jsonData, options: []),
    let repoArray = deserialised as? [[String: Any]],
    let firstRepo = repoArray.first,
    let ownerDictionary = firstRepo["owner"] as? [String: Any],
```

```
        let username = ownerDictionary["login"] as? String else {
    return
}
print(username)
```

That's a lot of optional unwrapping and casting just to get one value! Swift's strongly typed nature doesn't work well with JSON's loosely defined schema, which is why you have to do a lot of work to turn loosely typed information into strongly typed values.

Fortunately, there's a built-in solution that simplifies serialization and deserialization JSON: Codable. Using the Codable protocol, you can define a struct that conforms to the protocol, and then simply decode your data!

```
struct Owner : Codable {
    var userName: String
    // ...
}

let jsonData = // returned from the network guard
if let jsonOwners = try? JSONDecoder().decode(Owner.self, from:
jsonData) {
    print(jsonOwners.userName)
}
```

Working with XML

XML stands for **eXtensible Markup Language** and is a popular way of representing data for storage and transfer across a network. XML is a very flexible format and is used to represent many types of data. The current specification of HTML, which powers most of the web, is an implementation of XML.

The version of XML that we will concern ourselves with in this recipe is **RSS**, which stands for **Really Simple Syndication**. RSS is used to define a collection of time-ordered pieces of digestible content; these RSS feeds can then be used to aggregate content from a number of different sources. RSS is typically used as a distribution mechanism for news articles and podcasts.

In this recipe, we will learn how to read and write XML data by fetching and parsing the BBC News RSS feed.

Getting ready

The functionality to deal with XML data is provided by the Foundation framework. However, while the classes that help with reading XML data are available on all of Apple's platforms, the classes that assist with writing XML data are only available on the macOS platform.

This is an unfortunate oversight and means that if you need to write XML data within an iOS app, you will likely need to look for a third-party helper or build your own. We will investigate third-party helpers at the end of this recipe.

To investigate both reading and writing XML using the Foundation framework, we need to create a new macOS-based playground instead of an iOS-based playground, which we have been using so far in this book.

Create a new Swift playground as usual, but choose a **Blank** template from the **macOS** tab:

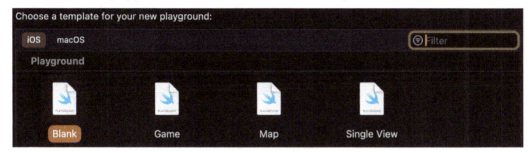

Figure 5.5 – Choosing a template

The RSS feed that we will retrieve and parse is from the front page of the BBC News website, which is http://feeds.bbci.co.uk/news/rss.xml.

Our first step is to retrieve the data at this URL so that we can start making sense of it. Since we previously covered retrieving information over the network, I'll add the code without further comment; check out the *Fetching data with URLSession* recipe in this chapter for more information:

```
import Foundation
import PlaygroundSupport
PlaygroundPage.current.needsIndefiniteExecution = true

func fetchBBCNewsRSSFeed() {
    let session = URLSession.shared
    let url = URL(string: "http://feeds.bbci.co.uk/news/rss.xml")!
    let dataTask = session.dataTask(with: url) { (data, response,
error) in
        guard let data = data, error == nil else {
            print(error ?? "Unexpected response")
            return
        }

        let dataAsString = String(data: data, encoding: .utf8)!
        print(dataAsString)
```

```
        }
        dataTask.resume()
    }


fetchBBCNewsRSSFeed()
```

When you run the playground, you will get an output that looks like the following:

```
<?xml version="1.0" encoding="UTF-8"?>
<?xml-stylesheet title="XSL_formatting" type="text/xsl" href="/shared/
bsp/xsl/rss/nolsol.xsl"?>
<rss xmlns:dc="http://purl.org/dc/elements/1.1/"
xmlns:content="http://purl.org/rss/1.0/modules/content/"
xmlns:atom="http://www.w3.org/2005/Atom" version="2.0"
xmlns:media="http://search.yahoo.com/mrss/">
<channel>
    <title><![CDATA[BBC News - Home]]></title>;
    <description><![CDATA[BBC News - Home]]></description>
    <link>https://www.bbc.co.uk/news/</link>
    <image>
        <url>https://news.bbcimg.co.uk/nol/shared/img/bbc_
news_120x60.gif</url>
        <title>BBC News - Home</title>
        <link>https://www.bbc.co.uk/news/</link>
    </image>
    <generator>RSS for Node</generator>
    <lastBuildDate>Sat, 15 Aug 2020 00:41:41 GMT</lastBuildDate>
    <copyright><![CDATA[Copyright: (C) British Broadcasting
Corporation, see http://news.bbc.co.uk/2/hi/help/rss/4498287.stm for
terms and conditions of reuse.]]></copyright>
    <language><![CDATA[en-gb]]></language>
    <ttl>15</ttl>
    <item>
        <title><![CDATA[Coronavirus: Thousands return to UK to beat
France quarantine]]></title>
        <description><![CDATA[Holidaymakers have just hours to
return to the UK to avoid the 14-day self-isolation requirement.]]></
description>
        <link>https://www.bbc.co.uk/news/uk-53782019</link>
        <guid isPermaLink="true">https://www.bbc.co.uk/news/uk-
53782019</guid>
        <pubDate>Fri, 14 Aug 2020 21:21:54 GMT</pubDate>
    </item>
    //... More items
</channel>
</rss>
```

How to do it...

The overall structure should be familiar to anyone who has seen HTML. Apart from the first two lines, which define the version and formatting of the XML, the information is structured with opening and closing tags. Consider the following example:

```
<link>https://www.bbc.co.uk/news/uk-53782019</link>
```

The name of the opening tag defines the content of this element of XML; in this case, it is a link. Then follows the content of the element, and the end of the content is defined by a closing tag that has a / character before its name.

In addition to this simple example, an XML element can have attributes that describe extra information about the content of the element:

```
<guid isPermaLink="true">https://www.bbc.co.uk/news/uk-53782019</guid>
```

These are defined as key-value pairs within the opening tag. The content of the XML element may be a string, as in the preceding examples, or it can be nested child XML elements:

```
<image>
     <url>http://news.bbcimg.co.uk/nol/shared/img/bbc_news_120x60.
gif</url>
     <title>BBC News - Home</title>
     <link>http://www.bbc.co.uk/news/</link>
</image>
```

Lastly, the content of an XML element can be data. This data might be represented as a string, especially if the string is likely to be longer, and may include line breaks, special characters, and other components that may be confused as being part of the enclosing XML formatting:

```
<title><![CDATA[Coronavirus: Thousands return to UK to beat France
quarantine]]></title>
```

Now that we have retrieved the XML, we want to parse it into something useful. The parser we will be using is provided by the Foundation framework and is available on iOS and macOS. It is called XMLParser. XMLParser is a **SAX parser**, which stands for **Simple API for XML**. The features of a SAX parser are as follows:

- Event-driven
- Low memory overhead
- Only retains relevant information
- One pass

The parser takes a delegate object that it will deliver event information to as it parses the document. It is the delegate object's responsibility to take and retain the relevant information from these delegate callbacks as the XML data is parsed, as the parser will not retain the parsed data.

We will step through a simple example to see how the parser reports events to the delegate. Here's the simple XML that we intend to parse:

```
<xml version="1.0" encoding="UTF-8"?>
<quotes>
    <quote attribution="Homer Simpson">
        Press any key to continue, where's the any key?
    <;/quote>
    <quote attribution="Unknown">
        Why do nerds confuse Halloween and Christmas? Because
OCT31=DEC25
    </quote>
</quotes>
```

The parser will start parsing the XML, character by character, and as an event is triggered, the delegate will be informed:

1. The first event will be the start of the document, where the parser will call this:

   ```
   func parserDidStartDocument(_ parser: XMLParser)
   ```

   Here, we can do any setup or resetting of the state that is required.

2. Then, the parser will move through the document until it reaches this point:

   ```
   <?xml version="1.0" encoding="UTF-8"?>
   <quotes>
       ** Parser is here **
       <quote attribution="Homer Simpson">
           Press any key to continue, where's the any key?
       </quote>
       <quote attribution="Unknown>
           Why do nerds confuse Halloween and      Christmas?
   Because OCT31=DEC25
       </quote>
   </quotes>
   ```

3. The parser has finished parsing the opening tag for the first element and so it fires the delegate callback:

   ```
   func parser(_ parser: XMLParser, didStartElement elementName:
   String, namespaceURI: String?, qualifiedName qName: String?,
   attributes attributeDict: [String : String] = [:]) {
       /*
   ```

```
    elementName = quotes
    namespaceURI = nil
    qName = nil
    attributeDict = [:]
    */
}
```

4. The parser then continues until it reaches this point:

```
<?xml version="1.0" encoding="UTF-8">
<quotes>
    <quote attribution="Homer Simpson">
        ** Parser is here **
        Press any key to continue, where's the any key?
    </quote>
    <quote attribution="Unknown">
        Why do nerds confuse Halloween and Christmas? Because
OCT31=DEC25
    </quote>
</quotes>
```

5. Since the parser has seen another starting tag, it fires the same delegate callback with information about this new element:

```
func parser(_ parser: XMLParser, didStartElement elementName:
String, namespaceURI: String?, qualifiedName qName: String?,
attributes attributeDict: [String : String] = [:]) {
    /*
    elementName = quote
    namespaceURI = nil
    qName = nil
    attributeDict = ["attribution": "Homer Simpson"]
    */
}
```

This time, as the element has attribute information, it is provided by the delegate callback in the `attributeDict` dictionary.

6. The parser now moves through the content of the first quote element. At some point, it fires the delegate callback with the content it has collected up to that point:

```
<?xml version="1.0" encoding="UTF-8"?>
<quotes>
    <quote attribution="Homer Simpson">
        Press any key to continue, ** Parser is here **where's
the any key?
    </quote>
```

```
    <quote attribution="Unknown>
        Why do nerds confuse Halloween and Christmas? Because
OCT31=DEC25
    </quote>
</quotes>
```

It then provides the content collected so far to the delegate:

```
func parser(_ parser: XMLParser, foundCharacters string: String)
{
    /*
    string = "Press any key to continue, "
    */
}
```

The reason the parser stops halfway through the content to fire the delegate callback is to make the most efficient use of memory. All the data that the parser processes must be kept in memory by the parser until it can be delivered to the delegate. Therefore, if the parser determines that memory usage is getting high, it will take the content it has collected so far and deliver it to the delegate. Once it has done this, it can free up the memory and start collecting further content afresh.

In this simple example, it is very unlikely that the parser will not provide all the content of the element in one delegate callback. It is, however, useful to see an example of this, as we have to account for the possibility, and it will affect how we implement the delegate later.

7. The parser will fire the same `foundCharacters` delegate callback until all of the content of an element has been delivered to the delegate:

```
<?xml version="1.0" encoding="UTF-8"?>
<quotes>
    <quote attribution="Homer Simpson">
        Press any key to continue, where's the any key?
        ** Parser is here **
    </quote>
    <quote attribution="Unknown">
        Why do nerds confuse Halloween and Christmas? Because
OCT31=DEC25
    </quote>
</quotes>
```

8. It then provides the new content since the last call to the delegate:

```
func parser(_ parser: XMLParser, foundCharacters string: String)
{
    /*
    string = "where's the any key?"
    */
}
```

9. The parser now processes the closing tag for the first quote element:

```
<?xml version="1.0" encoding="UTF-8"?>
<quotes>
    <quote attribution="Homer Simpson">
        Press any key to continue, where's the any key?
    </quote>
    ** Parser is here **
    <quote attribution="Unknown">
        Why do nerds confuse Halloween and Christmas? Because
OCT31=DEC25
    </quote>
</quotes>
```

10. Then, it fires the delegate callback, signaling the end of the element:

```
func parser(_ parser: XMLParser, didEndElement elementName:
String, namespaceURI: String?, qualifiedName qName: String?) {
    /*
    elementName = "quote"
    namespaceURI = nil
    qName = nil
    */
}
```

The parser will then continue to process the next quote element in the same way, firing the same sequence of `didStartElement`, followed by a number of `foundCharacters` callbacks, and finishing with a call to `didEndElement`.

11. Having finished processing the last quote element, the parser will process the closing tag of the `quotes` element:

```
<?xml version="1.0" encoding="UTF-8"?>
<quotes>
    <quote attribution="Homer Simpson">
        Press any key to continue, where's the any key?
    </quote>
    <quote attribution="Unknown">
        Why do nerds confuse Halloween and Christmas? Because
OCT31=DEC25
    </quote>
</quotes>
** Parser is here **
```

It will fire another `didEndElement` callback for the `quotes` element:

```
func parser(_ parser: XMLParser, didEndElement elementName:
String, namespaceURI: String?, qualifiedName qName: String?) {
    /*
    elementName = "quotes"
    namespaceURI = nil
    qName = nil
    */

}
```

12. Finally, the parser will fire a delegate callback to indicate that the parsing of the document is complete:

```
func parserDidEndDocument(_ parser: XMLParser) {


}
```

Now that you understand how the parser passes information to the delegate, we can return to our RSS example.

How it works...

You will remember that we retrieved XML data that looks like this:

```
<?xml version="1.0" encoding="UTF-8"?>
<?xml-stylesheet title="XSL_formatting" type="text/xsl" href="/shared/
bsp/xsl/rss/nolsol.xsl"?>
<rss xmlns:dc="http://purl.org/dc/elements/1.1/"
xmlns:content="http://purl.org/rss/1.0/modules/content/"
xmlns:atom="http://www.w3.org/2005/Atom" version="2.0"
xmlns:media="http://search.yahoo.com/mrss/">
<channel>
    <title><![CDATA[BBC News - Home]]></title>
        <description><![CDATA[BBC News - Home]]></description>
    <link>https://www.bbc.co.uk/news/</link>
    <image>
        <url>https://news.bbcimg.co.uk/nol/shared/img/ bbc_
news_120x60.gif</url>
        <title>BBC News - Home</title>
        <link>https://www.bbc.co.uk/news/</link>
    </image>
    <generator>RSS for Node</generator>
    <lastBuildDate>Sat, 15 Aug 2020 00:41:41 GMT</lastBuildDate>
    <copyright><![CDATA[Copyright: (C) British Broadcasting
Corporation, see http://news.bbc.co.uk/2/hi/help/rss/
4498287.stm for terms and conditions of reuse.]]></copyright>
```

```
<language><![CDATA[en-gb]]></language>
<ttl>15</ttl>
<item>
        <title><![CDATA[Coronavirus: Thousands return to UK to beat
France quarantine]]></title>
        <description><![CDATA[Holidaymakers have just hours to
return to the UK to avoid the 14-day self-isolation requirement.]]></
description>
        <link>https://www.bbc.co.uk/news/uk-53782019</link>
        <guid isPermaLink="true">https://www.bbc.co.uk/news/uk-
53782019</guid>
        <pubDate>Fri, 14 Aug 2020 21:21:54 GMT</pubDate>
</item>
//... More items
</channel>
</rss>
```

From this, we want to extract the news articles in a usable form, so let's define a `NewsArticle` model containing some useful information and place it near the top of the playground:

```
struct NewsArticle {
    let title: String
    let url: URL
}
```

Since the information we require will be spread over multiple delegate callbacks, our delegate will need to keep track of the information it has received, so it can be pieced together at the appropriate time.

Let's create a class object to be the delegate for the parser and have it conform to `XMLParserDelegate`:

```
class RSSNewsArticleBuilder: NSObject, XMLParserDelegate {

}
```

In the preceding XML, each news article is contained in an item element, so our delegate will need to keep track of when the parser is delivering content for the item element so that it can ignore content from other elements:

```
class RSSNewsArticleBuilder: NSObject, XMLParserDelegate {
    var inItem = false
    func parser(_ parser: XMLParser, didStartElement elementName:
String, namespaceURI: String?, qualifiedName qName: String?,
attributes attributeDict: [String : String] = [:]) {
        switch elementName {
        case "item":
            inItem = true
```

```
            default:
                break
            }
    }

    func parser(_ parser: XMLParser, didEndElement elementName:
String, namespaceURI: String?, qualifiedName qName: String?) {
            switch elementName {
            case "item":
                inItem = false
            default:
                break
            }
        }
    }
}
```

The two parts we want to extract from the item element to create our NewsArticle are the title and the URL. As we can see from the XML, the title is contained in a CDATA wrapper within a title element, and the URL is within a link element:

```
<item>
    <title><![CDATA[Coronavirus: Thousands return to UK to beat
France quarantine]]></title>
    <description><![CDATA[Holidaymakers have just hours to return
to the UK to avoid the 14-day self-isolation requirement.]]></
description>
    <link>https://www.bbc.co.uk/news/uk-53782019</link>
    <guid isPermaLink="true">https://www.bbc.co.uk/news/uk-53782019</
guid>
    <pubDate>Fri, 14 Aug 2020 21:21:54 GMT</pubDate>
</item>
```

We will, therefore, also need to keep track of when the parser is in the link element, and while it is within the link element, append the received content to a String property.

Similarly, we need to keep track of when the parser is in the title element, and when it is, append the received content to a Data property.

Let's add the extra properties we need to our RSSNewsArticleBuilder object:

```
class RSSNewsArticleBuilder: NSObject, XMLParserDelegate {
    var inItem = false
    var inTitle = false
    var inLink = false
    var titleData: Data?
```

```
        var linkString: String?
        //...
}
```

In the `didStartElement` method, we can check for these new element names we need to track. We must also remember to reset the link and title properties as we start the relevant element. This way, we don't continue to append content meant for the next item element to content from the previous one:

```
func parser(_ parser: XMLParser, didStartElement elementName: String,
namespaceURI: String?, qualifiedName qName: String?, attributes
attributeDict: [String : String] = [:]) {
    switch elementName {
    case "item":
        inItem = true
    case "title":
        inTitle = true
    titleData = Data()
    case "link":
        inLink = true
        linkString = ""
    default:
        break
    }
}
```

Now that we know when we are in the right elements, we can implement two of the

`XMLParserDelegate` methods to receive the relevant content and store it:

```
class RSSNewsArticleBuilder: NSObject, XMLParserDelegate {
    //...
    func parser(_ parser: XMLParser, foundCDATA CDATABlock: Data) {
        if inTitle {
            titleData?.append(CDATABlock)
        }
    }
    func parser(_ parser: XMLParser, foundCharacters string: String)
{
        if inLink {
            linkString?.append(string)
        }
    }
}
```

In the `didEndElement` method, we need to update our new properties and we can print out the values we have retrieved from the XML:

```swift
class RSSNewsArticleBuilder: NSObject, XMLParserDelegate {
    //...
    func parser(_ parser: XMLParser, didEndElement elementName:
String, namespaceURI: String?, qualifiedName qName: String?) {
        switch elementName {
        case "item":
            inItem = false
            guard let titleData = titleData,
                let titleString = String(data: titleData,
encoding: .utf8),
                let linkString = linkString,
                let link = URL(string: linkString)
            else {
                break
            }
            print(titleString) print(link)
        case "title":
            inTitle = false
        case "link":
            inLink = false
        default:
            break
        }
    }
    //...
}
```

Now that we have extracted the title and URL of the news article, we can use this to create a `NewsArticle` model object. First, let's create an array to hold the `NewsArticle` objects we will be creating:

```swift
class RSSNewsArticleBuilder: NSObject, XMLParserDelegate {
    var inItem = false
    var inTitle = false
    var inLink = false
    var titleData: Data?
    var linkString: String?
    var articles = [NewsArticle]()
    //...
}
```

We can create the `NewsArticle` object at the end of the item element as this is when we will have all the relevant content:

```
class RSSNewsArticleBuilder: NSObject, XMLParserDelegate {
    //...
    func parser(_ parser: XMLParser, didEndElement elementName:
String, namespaceURI: String?, qualifiedName qName: String?) {
        switch elementName {
        case "item":
            inItem = false
            guard let titleData = titleData,
                let titleString = String(data: titleData,
encoding:.utf8),
                let linkString = linkString,
                let link = URL(string: linkString)
            else {
                break
            }
            let article = NewsArticle(title: titleString, url:
link)
            articles.append(article)
        case "title":
            inTitle = false
        case "link":
            inLink = false
        default:
            break
        }
    }
    //...
}
```

Lastly, when the document starts, we should ensure that all the properties are reset:

```
class RSSNewsArticleBuilder: NSObject, XMLParserDelegate {
    //...
    func parserDidStartDocument(_ parser: XMLParser) {
        inItem = false
        inTitle = false
        inLink = false
        titleData = nil
        linkString = nil
        articles = [NewsArticle]()
    }
    //...
}
```

Now that we have completed the parser delegate, let's go back to our `fetchBBCNewsRSSFeed` function:

```swift
func fetchBBCNewsRSSFeed() {
    let session = URLSession.shared
    let url = URL(string: "http://feeds.bbci.co.uk/news/rss.xml")!
    let dataTask = session.dataTask(with: url) { (data, response,
error) in
        guard let data = data, error == nil else {
            print(error ?? "Unexpected response")
            return
        }
        let dataAsString = String(data: data, encoding: .utf8)!
        print(dataAsString)
    }
    dataTask.resume()
}
```

Once the XML data has been retrieved, we'll pass it to `XMLParser`, set up the delegate, and tell the parser to parse the data:

```swift
func fetchBBCNewsRSSFeed() {
    let session = URLSession.shared
    let url = URL(string: "http://feeds.bbci.co.uk/news/rss.xml")!
    let dataTask = session.dataTask(with: url) { (data, response,
error) in
        guard let data = data, error == nil else {
            print(error ?? "Unexpected response")
            return
        }
        let parser = XMLParser(data: data)
        let articleBuilder = RSSNewsArticleBuilder()
        parser.delegate = articleBuilder
        parser.parse()
        let articles = articleBuilder.articles
        print(articles)
    }
    dataTask.resume()
}
```

We want to provide the articles as an output from this function, so we can add a completion handler to provide an array of news articles or an error:

```swift
func fetchBBCNewsRSSFeed(completion: @escaping ([NewsArticle]?,
Error?) -> Void) {
    let session = URLSession.shared
    let url = URL(string: "http://feeds.bbci.co.uk/news/rss.xml")!
```

```
        let dataTask = session.dataTask(with: url) { (data, response,
error) in
            guard let data = data, error == nil else {
                completion(nil, error)
                return
            }
            let parser = XMLParser(data: data)
            let articleBuilder = RSSNewsArticleBuilder()
            parser.delegate = articleBuilder
            parser.parse()
            let articles = articleBuilder.articles
            completion(articles, nil)
        }
        dataTask.resume()
}
```

Finally, we can call this function, which will retrieve the RSS feed, parse it, and return an array of news articles:

```
fetchBBCNewsRSSFeed() { (articles, error) in
    if let articles = articles {
        print(articles)
    } else if let error = error {
        print(error)
    }
}
```

There's more...

Foundation also provides the ability to write XML data, although currently, this functionality is only available on macOS.

Having retrieved the RSS feed and created our news articles, let's write this information to an XML data structure and save it to disk. This XML will take the following form:

```
<articles>
    <article>
        <title>Donald Trump calls Fidel Castro 'brutal dictator'</
title>
        <url>http://www.bbc.co.uk/news/world-latin-america-
38118739</url>
    </article>
    <article>
        <title>Fidel Castro: Jeremy Corbyn praises 'huge figure'</
title>
```

```
            <url>http://www.bbc.co.uk/news/uk-38117068</url>
        </article>
    </articles>
```

At the root of the XML structure is an `articles` element, which contains multiple `article` elements, which in turn contain a `title` element and a `url` element.

To write the XML data, we will recreate the preceding structure using the `XMLDocument` and `XMLElement` objects. Once constructed, the `xmlData` property of the `XMLDocument` object provides the document as data.

Let's create a function to produce XML data from an array of `NewsArticle`:

```
func createXML(representing articles: [NewsArticle]) -> Data {
    let root = XMLElement(name: "articles")
    let document = XMLDocument(rootElement: root)
    for article in articles {
        let articleElement = XMLElement(name: "article")
        let titleElement = XMLElement(name: "title", stringValue:
article.title)
        let urlElement = XMLElement(name: "url", stringValue:
article.url.absoluteString)
        articleElement.addChild(titleElement)
        articleElement.addChild(urlElement)
        root.addChild(articleElement)
    }
    print(document.xmlString)
    return document.xmlData
}
```

We create each `XMLElement` and add it as a child to the element that we want to nest it within.

If you are building this in a storyboard, ensure that you place this function after `RSSNewsArticleBuilder`, and before the code that calls `fetchBBCNewsRSSFeed`, as this function will need to be available to the completion handler soon.

Our call to `fetchBBCNewsRSSFeed` will provide an array of `NewsArticle`, so we can pass this to our new function to write this information to XML data:

```
fetchBBCNewsRSSFeed() { (articles, error) in
    if let articles = articles {
        let articleXMLData = createXML(representing: articles)
        print(articleXMLData.length)
    } else if let error = error {
        print(error)
    }
}
```

Now that we have the data, we can obtain a URL for the `documents` directory, append the name of the file we will create, and write it to disk:

```
fetchBBCNewsRSSFeed() { (articles, error) in
    if let articles = articles {
        let xmlData = createXML(representing: articles)
        let documentsURL = FileManager.default.urls(for:
.documentDirectory, in: .userDomainMask).first!
        let writeURL = documentsURL.
appendingPathComponent("articles.xml")
        print("Writing data to: \(writeURL)")
        try! xmlData.write(to: writeURL)
    } else if let error = error {
        print(error)
    }
}
```

We have now retrieved an RSS feed, extracted useful information from it, written that information to a custom XML format, and saved that data to disk. Give yourself a pat on the back!

See also

Further information about `XMLParser` can be found in Apple's Foundation reference at `http://swiftbook.link/docs/xmlparser`.

Other XML parsers are available, which may have advantages over Apple's, including being able to write XML on iOS. They are as follows:

- RaptureXML: `https://github.com/ZaBlanc/RaptureXML`

- TBXML: `https://github.com/71squared/TBXML`

6

Understanding Concurrency in Swift

As we continue to learn more about Swift, eventually we will begin to build complex projects. With that complexity comes a lot more code for our apps to think through and execute. Over time, this could mean that our apps begin to feel weighty, slow, and unresponsive at runtime. Concurrency in Swift allows us to increase the performance and responsiveness of our code by allowing different tasks to be run seemingly at the same time, when possible.

In this chapter, we will learn how we can perform asynchronous tasks using **Grand Central Dispatch** (**GCD**) through the Dispatch framework and the higher-level operations in the Foundation framework that are also built on GCD. We will also take a look at the more current **Async/Await** framework, which brings a more modern approach, look, and feel to concurrency in Swift.

By the end of this chapter, you will know how to unlock the performance potential of your apps through concurrency. Understanding the multithreaded environment available on all Apple platforms, and the ways we can leverage them, is vital to building a fast and responsive app.

In this chapter, we have the following recipes:

- Using Dispatch queues for concurrency
- Leveraging `DispatchGroups`
- Implementing the operation class
- Async/Await in Swift

Technical requirements

All the code for this chapter can be found in this book's GitHub repository at `https://github.com/PacktPublishing/Swift-Cookbook-Third-Edition/tree/main/Chapter%206`.

Using Dispatch queues for concurrency

We live in a multicore computing world. Multicore processors are found in everything, from our laptops and mobile phones to our watches. With these multiple cores comes the ability to work in parallel. These concurrent streams of work are known as threads, and programming in a multithreaded way enables your code to make the best use of the processor's cores. Deciding how and when to create new threads and manage the available resources is a complex task, so Apple has built a framework to do the hard work for us. It is called Grand Central Dispatch, or GCD.

GCD handles thread maintenance and monitors the available resources while providing a simple, queue-based interface for getting concurrent work done. With the open sourcing of Swift, Apple also open sourced GCD in the form of `libdispatch` as Swift does not yet have built-in concurrency features.

In this recipe, we will explore some of the features of `libdispatch`, also known as the Dispatch framework, and see how we can use concurrency to build apps that are efficient and responsive.

Getting ready

We will see how we can improve the responsiveness of an app using GCD. So first, we need to start with an app that requires some improvement.

Go to `https://github.com/PacktPublishing/Swift-Cookbook-Third-Edition/tree/main/Chapter%206/PhotobookCreator_DispatchGroups`. Here, you will find the repository of an app that takes a collection of photos and turns them into a PDF photo book. You can download the app source files directly from GitHub or by using `git`:

```
git clone https://github.com/PacktPublishing/Swift-Cookbook-Third-
Edition/tree/main/Chapter%206/PhotobookCreator_DispatchGroups
```

If you build and run the app, you will see a collection of sample images, with the ability to add more:

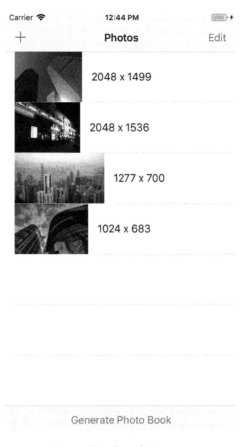

Figure 6.1 – Sample images

When you tap on **Generate Photo Book**, the app will take the photos you have chosen, resize them to the same size, and save them as a multi-page PDF that can then be exported or shared. Depending on how many photos are included and the performance of the device, this process can take a little time to complete. During this time, the whole interface is unresponsive; for example, you can't scroll through the pictures.

How to do it...

Let's examine why the app is unresponsive during photo book generation and how we can fix this:

1. Open up the `PhotoBookCreator` project and navigate to `PhotoCollectionView-Controller.swift`. In this file, you will find the following method:

    ```
    func generatePhotoBook(with photos: [UIImage]) {
        let resizer = PhotoResizer()
        let builder = PhotoBookBuilder()
    ```

```
        // Scale down (can take a while)
        var photosForBook = resizer.scaleToSmallest(of: photos)
        // Crop (can take a while)
        photosForBook = resizer.cropToSmallest(of: photosForBook)
        // Generate PDF (can take a while)
        let photobookURL = builder.buildPhotobook(with:
photosForBook)
        let previewController =
UIDocumentInteractionController(url: photobookURL)
        previewController.delegate = self
        previewController.presentPreview(animated: true)
    }
```

In this method, we call three functions that can take quite a long time to complete. We take the output of one function and feed it into the next function, and the result is a URL for our photo book, which we then launch with some UI to preview and export.

This work to resize and crop the photos, and then generate the photo book, is taking place in the same queue where UI touch events are processed, the main queue, which is why our UI is unresponsive.

2. To free up the main queue for UI events, we can create our own private queue, which we can use to execute our long-running functions:

```
import Dispatch

class PhotoCollectionViewController: UIViewController {
    //...
    let processingQueue = DispatchQueue(label: "Photo
processing queue")

    func generatePhotoBook(with photos: [UIImage]) {
        processingQueue.async { [weak self] in
            let resizer = PhotoResizer()
            let builder = PhotoBookBuilder()
            // Get smallest common size
            let size = resizer.smallestCommonSize(for:
photos)
            // Scale down (can take a while)
            var photosForBook = resizer.
scaleWithAspectFill(photos, to: size)
            // Crop (can take a while)
            photosForBook = resizer.centerCrop(photosForBook,
to: size)
            // Generate PDF (can take a while)
            let pbURL = builder.buildPhotobook(with:
photosForBook)
```

```
                // Show preview with export options
                let previewController =
    UIDocumentInteractionController(url: pbURL)
                previewController.delegate = self
                previewController.presentPreview(
    animated: true)
            }
        }
    }
```

By calling the async method on our `DispatchQueue` and providing a block of code, we are scheduling that block to be executed. GCD will execute that block when resources are available. Now, our long-running code isn't blocking the main queue, so our UI will remain responsive; however, if you were to run the app with just this change, you would get some very odd behavior when the app tries to show the preview view controller.

We just discussed the fact that UI touch events are delivered to the main queue, which is why we wanted to avoid blocking it; however, `UIKit` expects all UI events to happen on the main queue. Since we are currently creating and presenting the preview view controller from our private queue, we are defying this `UIKit` expectation, which can produce a number of bugs, including UI elements that never appear, or appear long after they were presented.

3. To solve this problem, we need to ensure that when we are ready to present our UI, we do that operation on the main queue:

```
func generatePhotoBook(with photos: [UIImage], using builder:
PhotoBookBuilder) {
    processingQueue.async { [weak self] in
        let resizer = PhotoResizer()
        let builder = PhotoBookBuilder()
        // Get smallest common size
        let size = resizer.smallestCommonSize(for: photos)
        // Scale down (can take a while)
        var photosForBook = resizer.
scaleWithAspectFill(photos, to: size)
        // Crop (can take a while)
        photosForBook = resizer.centerCrop(photosForBook, to:
size)
        // Generate PDF (can take a while)
        let pbURL = builder.buildPhotobook(with:
photosForBook)
        DispatchQueue.main.async {
            // Show preview with export options
            let previewController =
UIDocumentInteractionController (url: pbURL)
            previewController.delegate = self
            previewController.
```

```
presentPreview(            animated: true)
            }
        }
    }
```

Now, if you run the app, you will find that you can generate a photo book while still being able to interact with the UI, for instance, being able to scroll the table view.

How it works...

GCD uses queues to manage blocks of work in a multithreaded environment. Queues operate on a **first in first out** (FIFO) policy. When GCD determines that resources are available, it will take the next block from the queue and execute it. Once the block has finished executing, it will be removed from the queue:

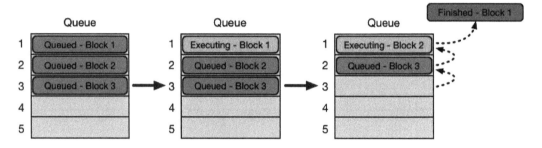

Figure 6.2 – FIFO policy

There are two types of `DispatchQueue`: serial and concurrent. With the simplest form of a queue, a serial queue, GCD will only execute one block at a time from the top of the queue. When each block finishes executing, it is removed from the queue, and each block moves up one position.

The main queue, which processes all UI events, is an example of a serial queue, and this explains why performing a long-running operation on the main queue will cause your UI to become unresponsive. While your long-running operation is executing, nothing else on the main queue will be executed until the long-running operation has finished.

With the second type of queue, a concurrent queue, GCD will execute as many blocks on different threads as resources allow. The next block to execute will be the block closest to the top of the stack that isn't already executing, and blocks are removed from the stack when finished:

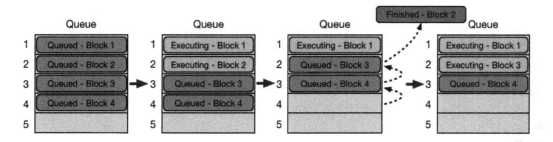

Figure 6.3 – Execution when the second type of queue is added

Concurrent queues can be useful when you have numerous operations that are independent of each other. We will look into concurrent queues further in the *Leveraging DispatchGroups* recipe.

See also

- The GitHub repository for `libdispatch`: `https://github.com/apple/swift-corelibs-libdispatch`

- Documentation for dispatch queues: `http://swiftbook.link/docs/dispatchqueue`

Leveraging DispatchGroups

In the previous recipe, we looked into using a private serial queue to keep our app responsive by moving long-running operations off the main queue. In this recipe, we will break our operations down into smaller, independent blocks and place them on a concurrent queue.

Getting ready

We are going to build on the app we improved in the last recipe, which is an app that will produce a PDF photo book from a collection of photos. You can get the code for this app at `https://github.com/PacktPublishing/Swift-Cookbook-Third-Edition/tree/main/Chapter%206` and choose the `PhotobookCreator_DispatchGroups` folder.

Open the project in Xcode and navigate to the `PhotoCollectionViewController.swift` file.

How to do it...

In the last recipe, we saw how dispatch queues operate on a FIFO policy. GCD will execute a block from the top of the queue and remove it from the queue when it has finished executing. The number of blocks that GCD will allow to execute at the same time will depend on the type of queue being used. Serial queues will only have one block of code being executed at any time; other blocks in the queues will have to wait until the block at the top of the queue has finished executing. However, for a concurrent queue, GCD will concurrently execute as many blocks as there are resources available. We can make more efficient use of a concurrent queue by breaking down the work into smaller, independent blocks, allowing them to be executed concurrently.

Take a look at the current implementation of the `generatePhotoBook` method. The only thing that has changed since the last recipe is that we now present the preview UI within a completion that is passed to the `generatePhotoBook` method. This simplifies the method and prevents us from needing to weakly capture `self` within the `async` block:

```
func generatePhotoBook(with photos: [UIImage], completion: @escaping
(URL) -> Void) {
    processingQueue.async {
        let resizer = PhotoResizer()
        let builder = PhotoBookBuilder()
        // Get smallest common size
        let size = resizer.smallestCommonSize(for: photos)
        // Scale down (can take a while)
        var photosForBook = resizer.scaleWithAspectFill(photos, to:
size)
        // Crop (can take a while)
        photosForBook = resizer.centerCrop(photosForBook, to: size)
        // Generate PDF (can take a while)
        let photobookURL = builder.buildPhotobook(with:
photosForBook)
        DispatchQueue.main.async {
            // Fire completion handler which will show the preview
UI
            completion(photobookURL)
        }
    }
}
```

The work we are doing is in one block of code that we place in a queue. Let's see whether we can break this down into smaller, independent pieces of work that can be executed concurrently. We can't perform the scale and crop operations concurrently, as they will be operating on the same `UIImage` objects, and we will not get the intended result if the image is cropped before it's scaled.

However, we can apply the scale and crop operation to each photo separately and perform that operation concurrently on the other photos. Once each photo has been scaled and cropped, we can use the processed images to generate the photo book:

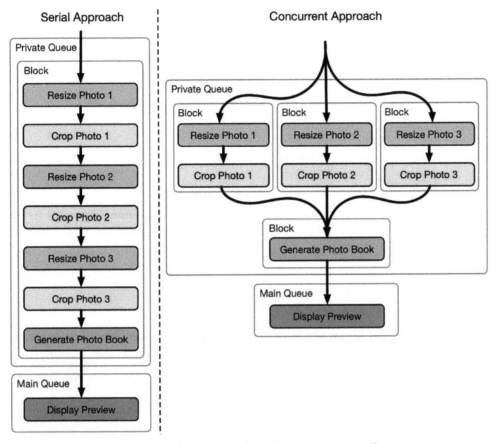

Figure 6.4 – Serial approach and concurrent approach

> **Note**
>
> Splitting the work up in this way may not make the overall operation faster, as there is an overhead to each block of work. The efficiency improvement of dividing the work into concurrent blocks will depend on the operation involved, and how many concurrent operations can run.

We now have blocks of work that can run concurrently, but we have given ourselves a new problem; how do we coordinate all these concurrent pieces of work so that we know they are all completed and we can start generating the photo book? Here, GCD can help us. We can use DispatchGroup to coordinate our operations on each of the images and be notified when they are all completed.

A dispatch group is like a turnstile at a stadium. Every time someone enters the stadium, they pass through the turnstile, and one extra person is counted as being in the stadium; then, at the end of the day, as people leave the stadium and pass through the turnstile, the number of people in the stadium decreases. Once there is no one left in the stadium, the lights can be turned off.

Let's use a dispatch group to coordinate the work of our photo book creator:

1. First, we will create a dispatch group:

    ```
    let group = DispatchGroup()
    ```

2. Every time we start a block of work to resize a photo, we will enter the group:

    ```
    group.enter()
    ```

3. Once the work is finished, we will leave the group:

    ```
    group.leave()
    ```

4. Finally, we will ask the group to notify us when the last resize operation has finished and left the group. Then, we can take the processed files and generate the photo book:

    ```
    group.notify(queue: processingQueue) {
        //.. generate photo book
        //.. execute completion handler
    }
    ```

5. Let's take a look at our `generatePhotoBook` method, now using a concurrent queue and dispatch groups:

    ```
    let processingQueue = DispatchQueue(label: "Photo processing
    queue", attributes: .concurrent)

    func generatePhotoBook(with photos: [UIImage], completion: @
    escaping
    (URL) -> Void) {
        let resizer = PhotoResizer()
        let builder = PhotoBookBuilder()
        // Get smallest common size
        let size = resizer.smallestCommonSize(for: photos)
        let processedPhotos = NSMutableArray(array: photos)
        let group = DispatchGroup()
        for (index, photo) in photos.enumerated() {
            group.enter()
            processingQueue.async {
                // Scale down (can take a while)
                var photosForBook = resizer.
    scaleWithAspectFill([photo], to: size)
                // Crop (can take a while)
                photosForBook = resizer.centerCrop([photo], to:
    size)
    ```

```
                        // Replace original photo with processed photo
                        processedPhotos[index] = photosForBook[0]
                        group.leave()
                    }
                }
        group.notify(queue: processingQueue) {
                guard let photos = processedPhotos as? [UIImage] else
    {
                    return
                }
                // Generate PDF (can take a while)
                let photobookURL = builder.buildPhotobook(with:
    photos)
                DispatchQueue.main.async {
                    completion(photobookURL)
                }
            }
        }
    }
```

How it works...

Dispatch queues are serial by default, so to create a concurrent queue instead, we pass the .concurrent attribute when it is created:

```
let processingQueue = DispatchQueue(label: "Photo processing queue",
attributes: .concurrent)
```

Before we loop through all the photos, we set up anything that isn't specific to each photo:

```
let resizer = PhotoResizer()
let builder = PhotoBookBuilder()
// Get smallest common size
let size = resizer.smallestCommonSize(for: photos)
let processedPhotos = NSMutableArray(array: photos)
let group = DispatchGroup()
```

This includes creating DispatchGroup, which we will use to coordinate the work. Since our photo resizing will now be happening concurrently, we need a place to collect the photos once they have been processed. We can use a Swift array for this; however, a Swift array is a value type, so we can't use it from within multiple blocks, as each block will be taking a copy of the array, not the original array itself.

To solve this with a Swift array, we would need to make the `processedPhotos` array property on the view controller, which would require us to weakly capture self in the blocks and later unwrap the `processedPhotos` array to access it. A simpler way to solve this problem is to use a collection that has reference semantics; the Foundation framework provides that in the form of `NSArray` and `NSMutableArray`. As we saw earlier in this chapter, it's important to understand the semantics of the construct being used and pick the right tool for the right job:

```swift
for (index, photo) in photos.enumerated() {
    group.enter()
    processingQueue.async {
        // Scale down (can take a while)
        var photosForBook = resizer.scaleWithAspectFill([photo], to:
size)
        // Crop (can take a while)
        photosForBook = resizer.centerCrop([photo], to: size)
        // Replace original photo with processed photo
        processedPhotos[index] = photosForBook[0]
        group.leave()
    }
}
```

For each photo, we enter the group and place the resize work on the concurrent queue. We can use the same scale and crop methods that we used previously, just passing an array containing one photo. Once the work is completed, we'll replace the original photo with the processed photo in the array and leave the group.

Once every block has left the group, this `notify` block will execute. We retrieve the processed photos and use them to generate the photo book. Finally, we ensure that the completion handler is executed on the main queue:

```swift
group.notify(queue: processingQueue) {
    guard let photos = processedPhotos as? [UIImage] else {
        return
    }
    // Generate PDF (can take a while)
    let photobookURL = builder.buildPhotobook(with: photos)
    DispatchQueue.main.async {
        completion(photobookURL)
    }
}
```

If you build and run the app, you can still generate a photo book and the UI is still responsive, and now GCD can make the best use of the available resources to generate our photo book.

See also

- Documentation relating to dispatch queues: `https://developer.apple.com/documentation/dispatch/dispatchqueue`
- Documentation relating to dispatch groups: `https://developer.apple.com/documentation/dispatch/dispatchgroup`

Implementing the operation class

In this chapter so far, we have taken our long-running operations and scheduled them as blocks of code, called closures, on dispatch queues. This has made it really easy to move long-running code off of the main queue, but if we intend to reuse this long-running code, pass it around, track its state, and generally deal with it in an object-orientated way, a closure is not ideal.

To solve this, the Foundation framework provides an object, `Operation`, that allows us to wrap up our block of work within an encapsulated object.

In this recipe, we will take the photo book app we used throughout this chapter and convert our long-running blocks into an `Operation` instance.

Getting ready

We are going to build on the app we improved in the last recipe, which is an app that will produce a PDF photo book from a collection of photos. You can get the code for this app at `https://github.com/PacktPublishing/Swift-Cookbook-Third-Edition/tree/main/Chapter%206` and choose the `PhotobookCreator_StartOperations` folder.

Open the folder and navigate to the `PhotoCollectionViewController.swift` file.

How to do it...

Let's recap how we broke the work down into independent parts:

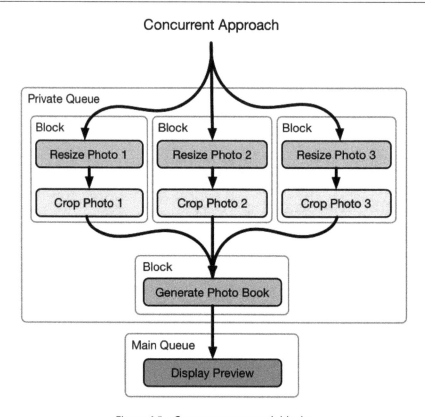

Figure 6.5 – Concurrent approach blocks

We can turn each of these blocks of work into separate operations:

1. Let's create an operation to scale and crop each photo.

2. We define an operation by sub-classing the Operation class, so in the project, create a new Swift file and call it PhotoResizeOperation.swift.

3. In the simplest Operation implementation, we only need to override one method, main(). So, let's copy and paste the relevant code from our generatePhotobook method. This main() method will be executed when the operation starts:

```
import UIKit

class PhotoResizeOperation: Operation {
    override func main() {
        // Scale down (can take a while)
        var photosForBook = resizer.
scaleWithAspectFill([photo], to: size)
        // Crop (can take a while)
        photosForBook = resizer.centerCrop([photo], to: size)
```

```
                    // Replace original photo with processed photo
                    processedPhotos[index] = photosForBook[0]
            }
    }
```

4. Copying and pasting the code is not enough, as there are a number of dependencies that were previously captured by the block. Now, we have to explicitly provide these dependencies to the operation:

```
class PhotoResizeOperation: Operation {
        let resizer: PhotoResizer
        let size: CGSize
        let photos: NSMutableArray
        let photoIndex: Int

        init(resizer: PhotoResizer, size: CGSize, photos:
NSMutableArray, photoIndex: Int) {

                self.resizer = resizer
                self.size = size
                self.photos = photos
                self.photoIndex = photoIndex
        }

        override func main() {
                // Retrieve the photo to be resized.
                guard let photo = photos[photoIndex] as? UIImage else
{

                        return
                }
                // Scale down (can take a while)
                var photosForBook = resizer.
scaleWithAspectFill([photo], to: size)
                // Crop (can take a while)
                photosForBook = resizer.centerCrop(photosForBook, to:
size)
                photos[photoIndex] = photosForBook[0]
        }
}
```

5. We have converted our resize block into an operation. We now need to do the same for the block that generates the photo book:

```
import UIKit

class GeneratePhotoBookOperation: Operation {
```

```
let builder: PhotoBookBuilder
let photos: NSMutableArray
var photobookURL: URL?

init(builder: PhotoBookBuilder, photos: NSMutableArray) {
    self.builder = builder
    self.photos = photos
}

override func main() {
    guard let photos = photos as? [UIImage] else {
        return
    }
    // Generate PDF (can take a while)
    photobookURL = builder.buildPhotobook(with: photos)
}
}
```

We pass the dependencies into the operation, just like in `PhotoResizeOperation`. The output of this operation is a URL for the resulting photo book. We expose that as a property in the operation so that it can be retrieved outside the operation.

6.   With our blocks of work converted into operations, let's switch over to `PhotoCollectionViewController.swift` and update our `generatePhotoBook` method to use this new operation:

```
let processingQueue = OperationQueue()

func generatePhotoBook(with photos: [UIImage], completion: @
escaping (URL) -> Void) {
    let resizer = PhotoResizer()
    let builder = PhotoBookBuilder()
    // Get smallest common size
    let size = resizer.smallestCommonSize(for: photos)
    let processedPhotos = NSMutableArray(array: photos)
    let generateBookOp = GeneratePhotoBookOperation(builder:
builder, photos: processedPhotos)
    for index in 0..<processedPhotos.count {
        let resizeOp = PhotoResizeOperation(resizer: resizer,
size: size, photos: processedPhotos, photoIndex: index)
        generateBookOp.addDependency(resizeOp)
        processingQueue.addOperation(resizeOp)
    }
    generateBookOp.completionBlock = { [weak generateBookOp] in
        guard let pbURL = generateBookOp?.photobookURL else {
```

```
            return
        }
        OperationQueue.main.addOperation {
            completion(pbURL)
        }
    }
    processingQueue.addOperation(generateBookOp)
}
```

Let's walk through the changes step by step:

1. Where we were previously using `DispatchQueue` to manage the execution of our blocks, operations are now managed with `OperationQueue`:

```
let processingQueue = OperationQueue()
```

2. The method signature in the following code and the dependencies we need to generate upfront remain the same:

```
func generatePhotoBook(with photos: [UIImage], completion: @
escaping (URL) -> Void) {
    let resizer = PhotoResizer()
    let builder = PhotoBookBuilder()
    // Get smallest common size
    let size = resizer.smallestCommonSize(for: photos)
    let processedPhotos = NSMutableArray(array: photos)
```

3. Next, we create the operation to generate the photo book, passing in the dependencies:

```
let generateBookOp = GeneratePhotoBookOperation(builder:
builder, photos: processedPhotos)
```

Although the operation will be executed last, we create it first so that we can make it dependent on the resize operations we are about to create. An operation does not execute immediately upon creation. It will only execute when the `start()` method of `Operation` is called, which can be called manually; or, if `Operation` is placed on `OperationQueue`, it will be called by the queue as appropriate:

```
for index in 0..<processedPhotos.count {
    let resizeOp = PhotoResizeOperation(resizer: resizer, size: size,
photos: processedPhotos, photoIndex: index)
    generateBookOp.addDependency(resizeOp)
    processingQueue.addOperation(resizeOp)
}
```

Now, as you can see from the preceding code, we loop through the number of photos that we intend to process and create a resize operation for each, passing in the dependencies.

With our move to use `Operation`, one thing we have lost is the use of `DispatchGroup`, which we used to ensure that we only generated the photo book once all the photo resize blocks had completed. We can, however, achieve the same goals using operation dependencies. An operation can be declared as dependent on a set of other operations, so it will not begin executing until the operations it depends on have finished. To ensure that the `generateBookOp` operation, which we just created, only executes when all the `PhotoResizeOperation` operations are complete, we add each of them as a dependency of `generateBookOp`.

With this done, we can place each `PhotoResizeOperation` in `OperationQueue`:

```
generateBookOp.completionBlock = { [weak generateBookOp] in
    guard let pbURL = generateBookOp?.photobookURL else {
        return
    }
    OperationQueue.main.addOperation {
        completion(pbURL)
    }
}
```

`Operation` has a `completionBlock` property; any block set here will be executed once the operation has completed. We can use this to fire our completion handler on the main queue. Since we need to provide the completion handler with the URL to the photo book created by `generateBookOp`, we can retrieve this from within the block, as we know that the operation will be finished and the URL will be there. However, we need to be careful.

We are providing a closure to `generateBookOp`, which will be retained, and we are using, and therefore capturing and retaining, the `generateBookOp` operation in the same block. This will lead to a retain cycle, and `generateBookOp` will never get released from memory. To avoid this retain cycle, we specify that we want to weakly capture `generateBookOp` in the block we provide, using the `[weak generateBookOp]` capture list. This won't increment the retain count, preventing the retain cycle from happening.

Much like `DispatchQueue`, `OperationQueue` has an available property that provides a reference to the main queue, upon which the UI events are processed. Also, `OperationQueue` has a convenience method that will take a block of code, bundle it into an `Operation` class, then add it to the queue. We use this to ensure that the completion handler is executed on the main queue:

```
processingQueue.addOperation(generateBookOp)
```

As the final step, we put the `generateBookOp` operation in the processing queue. It's important that we do this as the last step because once placed in the queue, the operation may be executed immediately, but we don't want it to be executed immediately. We only want `generateBookOp` to be executed once all the resize operations are complete, and if we place the operation on the queue before setting up the dependencies, this could happen.

Now that we have transitioned our app over to using `Operation`, let's build and then run and verify that everything works just as it did before.

Users of our photo book app currently do not have the ability to cancel the generation of a photo book once the process has started, so let's add that functionality:

1. We will examine our two operations and look for opportunities to check the `isCancelled` property and exit early. Switch to `PhotoResizeOperation.swift` and add `isCancelled` checks to the `main()` method:

```
override func main() {
    // Check if operation has been cancelled
    guard isCancelled == false else {
        return
    }
    guard let photo = photos[photoIndex] as? UIImage else {
        return
    }
    // Scale down (can take a while)
    var photosForBook = resizer.scaleWithAspectFill( [photo],
to: size)
    // Check if operation has been cancelled
    guard isCancelled == false else {
        return
    }
    // Crop (can take a while)
    photosForBook = resizer.centerCrop(photosForBook, to: size)
    photos[photoIndex] = photosForBook[0]
}
```

Before each piece of long-running work, we check the `isCancelled` property, and if it is `true`, we return early, which will finish the operation.

2. We can do the same in `GeneratePhotoBookOperation.swift`:

```
override func main() {
    // Check if operation has been cancelled
    guard isCancelled == false else {
        return
    }
    guard let photos = photos as? [UIImage] else {
        return
    }
    // Generate PDF (can take a while)
    photobookURL = builder.buildPhotobook(with: photos)
}
```

3. Next, as an exercise for the reader, we might want to add some UI that allows the user to cancel the photo book generation once it is in progress.

4. Once the user chooses to cancel generating a photo book, we can call the following command:

    ```
    processingQueue.cancelAllOperations()
    ```

    This will fire the `cancel()` method on all the operations in the queue.

We now have an app with a cancelable, long-running operation.

How it works...

How does `OperationQueue` know when to start an operation and when to remove it from the queue? It knows by monitoring the operation's state. The `Operation` class goes through a number of state transformations during its life cycle. The following diagram describes how these state transformations occur:

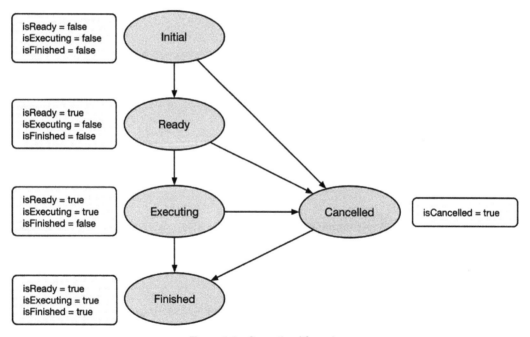

Figure 6.6 – Operation life cycle

Information about the operation's state is exposed through a number of Boolean properties on `Operation`, and the operation queue uses the properties to know when to perform certain actions on the operations. Let's look at these properties one by one:

```
var isReady: Bool
```

An operation will return `true` for `isReady` when all its dependencies are finished. If it doesn't have any dependencies, it will always return `true`. The queue will only start executing an operation if `isReady` is `true`:

```
var isExecuting: Bool
```

Once `start` is called on an operation, either manually or by a queue, `isExecuting` will return `true`, and when the operation has finished executing, `isExecuting` will revert to returning `false`.

Since operations remain on the queue until they have finished, the queue uses the `isExecuting` property to ensure that it doesn't call `start` on an operation that has already started:

```
var isFinished: Bool
```

Once the operation has finished doing whatever processing is required, `isFinished` should return `true`. When `isFinished` starts to return `true`, it will be removed from the queue, and the queue will no longer maintain a reference to the operation. For the simplest implementation of `Operation`, as we implemented earlier, `isFinished` returns `true` automatically when the `main()` method has finished executing:

```
var isCancelled: Bool
```

Operations can be canceled by calling the `cancel()` method on the operation. Once called, the `isCancelled` property will return `true`. This can be used to exit early from a long-running operation, but it is up to you to check the `isCancelled` method and interrupt any long-running code if it returns `true`.

See also

Documentation relating to the `Operation` class: `http://swiftbook.link/docs/operation`

Async/Await in Swift

Starting with Swift 5.5, we were introduced to yet another, though helpful, way to write and perform asynchronous code using **Async/Await**. For anyone who has worked in JavaScript or C# before, this may seem familiar and welcomed. What Async/Await does is allow us to write async functions just like we would any other synchronous code, and then call them using the `await` keyword.

In this recipe, we will take our `PhotobookCreator` app and swap out how we use Dispatch for Async/Await, highlighting some of the advantages it brings to Swift.

Getting ready

We will see how we can improve the responsiveness, readability, and safety of an app using Async/ Await, so we will work with an improved version of our app. Go to `https://github.com/ PacktPublishing/Swift-Cookbook-Third-Edition/tree/main/Chapter%206/ PhotobookCreator_AsyncAwait`. Here, you will find the repository of an app that takes a collection of photos and turns them into a PDF photo book, but using Async/Await. You can download the app source files directly from GitHub or by using `git`:

```
git clone    https://github.com/PacktPublishing/Swift-Cookbook-
Third-Edition/tree/main/Chapter%206/PhotobookCreator_AsyncAwait/
PhotobookCreator
```

How to do it...

While using Dispatch, we were able to achieve asynchronous code, but there were a few inconveniences. We had to create `DispatchQueue`, pass in a completion handler, and ensure that we returned to the main queue to run our completion.

As we'll see, Async/Await handles some of those details for us and makes much of what we're trying to achieve more readable:

1. First, in `PhotoCollectionViewController`, let's refactor `generatePhotoBook`. Instead of taking a closure, let's return what we would've passed into the handler: a URL. To signal that we'll be running this asynchronously, let's use the `async` keyword:

    ```
    func generatePhotoBook(with photos: [UIImage]) async -> URL {
        //...
    }
    ```

2. Now, we'll refactor the body of our function. Since `async` functions are made to look like synchronous functions, you'll notice it reads very straightforward. However, we'll add the `await` keyword in front of the calls we expect to be waiting on. Lastly, we'll simply return the URL we're looking for:

    ```
    func generatePhotoBook(with photos: [UIImage]) async -> URL {
        let resizer = PhotoResizer()
        let builder = PhotoBookBuilder()
            // Get smallest common size
        let size = await resizer.smallestCommonSize(for: photos)
        // Scale down (can take a while)
        var photosForBook = await resizer.
    scaleWithAspectFill(photos, to: size)
        // Crop (can take a while)
        photosForBook = await resizer.centerCrop(photosForBook, to:
    size)
    ```

```
        // Generate PDF (can take a while)
        let photobookURL = await builder.buildPhotobook(with:
    photosForBook)
        return photobookURL
        }
```

3. The compiler will start complaining that we're awaiting non-async functions. Let's fix this by simply adding the `async` keyword to each function we plan to call asynchronously:

```
func smallestCommonSize(for photos: [UIImage]) async -> CGSize

func scaleWithAspectFill(_ photos: [UIImage], to size: CGSize)
async -> [UIImage]

func centerCrop(_ photos: [UIImage], to size: CGSize) async ->
[UIImage]

func buildPhotobook(with photos: [UIImage]) async -> URL
```

4. Lastly, we update our caller, `generateButtonPressed`. Since it is `IBAction`, which is synchronous, we cannot use the `async` keyword. Instead, we'll wrap our asynchronous code using a `Task` closure:

```
@IBAction func generateButtonPressed(sender: UIBarButtonItem) {
        activityIndicator.startAnimating()
        Task {
                let photobookURL = await generatePhotoBook(with:
    photos)
            activityIndicator.stopAnimating()
            let previewController =
    UIDocumentInteractionController(url: photobookURL)
            previewController.delegate = self
            _ = previewController.presentPreview(
    animated: true)
        }
}
```

How it works...

A huge advantage of using Async/Await is that it allows us to clearly mark code that will be used asynchronously (using `async`) and that we must wait for before moving on (using `await`).

This is exceptionally helpful when looking at the body of `generatePhotoBook`. First, there's no need for any closure since the entire function has been marked as asynchronous. Next, we can clearly see the steps our code will be executing by taking note of where `await` appears. Lastly, we simply return the value our caller wants to use.

Returning a value instead of calling a closure (specifically on the main thread as we do using `Dispatch`) introduces not only better readability but also safety. It puts the responsibility of the callers' code back where it belongs: in the hands of the caller itself.

We achieve this in `generateBackButtonPressed`. Yes, we still end up using a closure, specifically a `Task` closure. However, we use it to simply signify that we want to execute an asynchronous call, as opposed to packaging up and sending off our code to be called and executed elsewhere:

```
Task {
    let photobookURL = await generatePhotoBook(with: photos)
    activityIndicator.stopAnimating()
    let previewController = UIDocumentInteractionController(url:
photobookURL)
    previewController.delegate = self
    _ = previewController.presentPreview(animated: true)
}
```

We simply await the call to `generatePhotoBook`, soundly trusting that once it completes execution, we can safely move forward with updating our UI with a URL.

The syntax changes between using Dispatch and Async/Await provide a dramatic difference in readability, simplicity, and safety, even in the simple use case we used them in. We've only scratched the surface of the capabilities offered by Async/Await.

See also

Documentation relating to Async/Await: `https://docs.swift.org/swift-book/documentation/the-swift-programming-language/concurrency/`

7

Building iOS Apps with UIKit

In the previous chapter, we got to play with a pre-made app so we could learn more about concurrency in Swift. Now, we'll be building our very own iOS app from scratch using Swift and the **Xcode** IDE. To do this, we'll use **UIKit** and storyboards, the more traditional way of building out an app UI. Once we've built our app, we'll look at how we can incorporate unit tests and UI tests.

By the end of this chapter, you will understand how to build a complete app and a UI so that anyone who uses it can trigger, interact with, and see the results of the Swift code that lies beneath.

In this chapter, we will cover the following recipes:

- Building an iOS app using UIKit and storyboards
- Unit and integration testing with XCTest
- UI testing with XCUITest

Technical requirements

For this chapter, you'll need the latest version of Xcode from the Mac App Store.

All the code for this chapter can be found in this book's GitHub repository at `https://github.com/PacktPublishing/Swift-Cookbook-Third-Edition/tree/main/Chapter%207/CocoaTouch`.

Building an iOS app using UIKit and storyboards

The focus of this book is ultimately on the Swift programming language itself, as opposed to the use of the language to produce apps for Apple platforms or to build server-side services. That being said, it can't be ignored that the vast majority of the Swift code being written is for building, or building upon, iOS and iPadOS apps.

In this recipe, we will take a brief look at how we can interact with Apple's Cocoa Touch frameworks using Swift and begin to build and create our very own iOS app.

Cocoa Touch is the name given to the collection of UI frameworks available as part of the iOS SDK. Its name derives from the Cocoa framework on macOS, which provides UI elements for macOS apps. While Cocoa on macOS is a framework in its own right, Cocoa Touch is a collection of frameworks that provide UI elements for iOS apps and handle the app's life cycle; the core of these frameworks is UIKit.

Getting ready

First, we'll need to create a new iOS app project:

1. From the Xcode menu, choose **File**, then **New**.

2. From the dialog box that opens, choose **App** from the **iOS** tab:

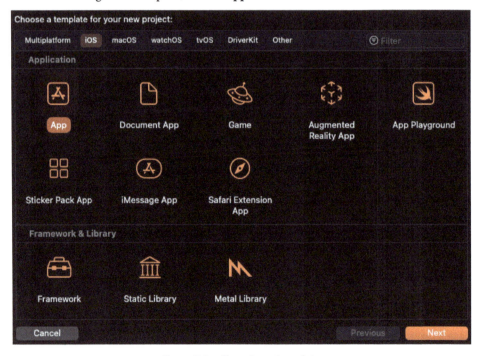

Figure 7.1 – Choosing a template

The next dialog box asks you to enter details about your app, pick a product name and organization name, and add an organization identifier in reverse DNS style.

Reverse DNS style means to take a website that you or your company owns and reverse the order of the domain name components. So, for example, `http://maps.google.com` becomes `com.google.maps`:

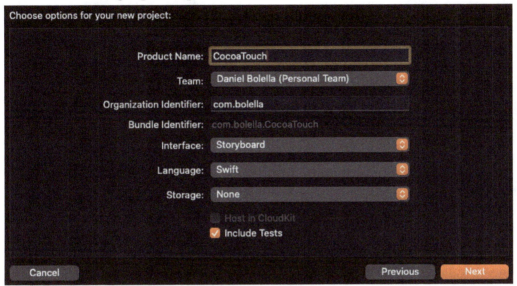

Figure 7.2 – Options for a new project

Pay attention to the preceding choices as not all of them may be selected by default. For this recipe, the ones that are important to us are **Interface** and **Include Tests**, both of which we'll cover later on in this chapter when we look at unit testing with **XCTest** and UI testing with **XCUITest**.

3. Once you've chosen a save location on your Mac, you will be presented with the following Xcode layout:

Figure 7.3 – New project template

Here, we have the start of our project – it's not much, but it's where all new iOS apps begin.

From this menu, press **Product | Run**. Xcode will now compile and run your app in a simulator.

How to do it...

Continuing from a previous recipe, we'll build our app based on data that is returned from the public GitHub API:

1. In the file explorer, click on `Main.storyboard`; this view is a representation of what the app will look like and is called **Interface Builder**. At the moment, there is only one blank screen visible, which matches what the app looked like when we ran it earlier. This screen represents a **view controller object**; as the name suggests, this is an object that controls views.

2. We will display our list of repositories in a table. We actually want to create a view controller class that is a subclass of `UITableViewController`. So, from the menu, choose **File**, then **New**, and select a **Cocoa Touch Class** template:

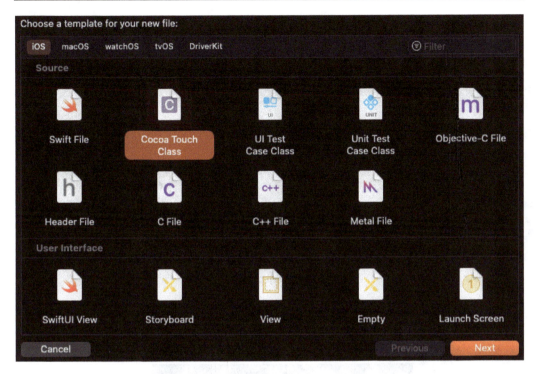

Figure 7.4 – New file template

3.   We will be displaying repositories in this view controller, so let's call it `ReposTableView-Controller`. Specify that it's a subclass of `UITableViewController` and ensure that the language is Swift:

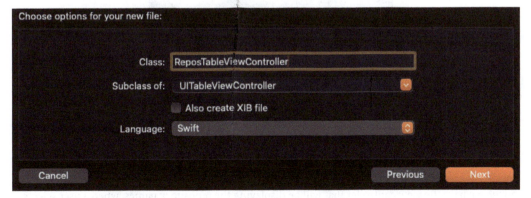

Figure 7.5 – New filename and subclass

Now that we have created our view controller class, let's switch back to `Main.storyboard` and delete the blank view controller that was created for us.

4. From the object library, find the `Table View Controller` option and drag it into the Interface Builder editor:

Figure 7.6 – Object library

5. Now that we have a table view controller, we want this controller to be part of our custom subclass. To do this, select the controller, go into the class inspector, enter `ReposTableViewController` as the `Class` type, and press *Enter*:

Figure 7.7 – Custom class inspector

Although we have the view controller that will be displaying the repository names, when a user selects a repository, we want to present a new view controller that will show details about that particular repository. We will cover what type of view controller that is and how we present it shortly, but first, we need a mechanism for navigating between view controllers.

If you have ever used an iOS app, you will be familiar with the standard *push* and *pop* way of navigating between views. The following screenshot shows an app in the middle of that transition:

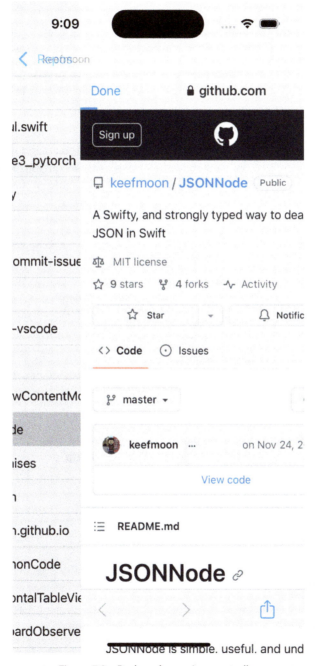

Figure 7.8 – Push and pop view controller

The management of these view controllers, as well as their presentation and dismissal transitions, is handled by a navigation controller, which is provided by Cocoa Touch in the form of `UINavigationController`. Let's take a look:

1. To place our view controller inside a navigation controller, select **ReposTableViewController** in Interface Builder. Then, from the Xcode menu, go to **Editor | Embed In | Navigation Controller**.

   This will add a navigation controller to the storyboard and set the selected view controller as its root view controller (if there is an existing view controller already inside the storyboard from the initial project we created, this can be highlighted and deleted).

2. Next, we need to define which view controller is initially on the screen when the app starts. Select **Navigation Controller** on the left-hand side of the screen and within the property inspector, select **Is Initial View Controller**. You will see that an entry arrow will point toward the navigation controller on the left, indicating that it will be shown initially.

3. With this set up, we can start working on our `ReposTableViewController` by selecting it from the **File** navigator menu.

   When we created our view controller, the template gave us a bunch of code, with some of it commented out. The first method that the template provides is `viewDidLoad`. This is part of a set of methods that cover the life cycle of the root view that the view controller is managing. Full details about the view life cycle and its relevant method calls can be found at https://developer.apple.com/documentation/uikit/view_controllers/displaying_and_managing_views_with_a_view_controller/#3370691.

   `viewDidLoad` is fired quite early on in the view controller's life cycle but before the view controller is visible to the user. Due to this, it is a good place to configure the view and retrieve any information that you want to present to the user.

4. Let's give the view controller a title:

   ```
   class ReposTableViewController: UITableViewController {
       override func viewDidLoad() {
           super.viewDidLoad()
           self.title = "Repos"
       }
       //...
   }
   ```

   Now, if you build and run the app, you'll see a navigation bar with the title we just added programmatically.

5. Next, we'll fetch and display a list of GitHub repositories. Implement the following snippet of code in order to fetch a list of repositories for a specific user:

   ```
   @discardableResult
   internal func fetchRepos(forUsername username: String,
   ```

```
completionHandler: @escaping (FetchReposResult) -> Void)->
URLSessionDataTask? {
    let urlString = "https://api.github.com/users/\ (username)/
repos"
    guard let url = URL(string: urlString) else {
        return nil
    }
    var request = URLRequest(url: url)
    request.setValue("application/vnd.github.v3+json",
forHTTPHeaderField: "Accept")
    let task = session.dataTask(with: request) { (data,
response, error) in
        // First unwrap the optional data
        guard let data = data else {
            let error = ResponseError. requestUnsuccessful
            completionHandler(.failure(error))
            return
        }
        do {
            let decoder = JSONDecoder()
            let responseObject = try decoder. decode([Repo].self,
from: data)
            completionHandler(.success(responseObject))
        } catch {
            completionHandler(.failure(error))
        }
    }
    task.resume()
    return task
}
```

6. Let's add the following highlighted code to the top of the file, before the start of the class definition. We will also add a session property to the view controller, which is needed for the network request:

```
import UIKit

struct Repo: Codable {
    let name: String?
    let url: URL?
    enum CodingKeys: String, CodingKey {
        case name = "name"

        case url = "html_url"
    }
}
```

```
    }

    enum FetchReposResult {
        case success([Repo])
        case failure(Error)
    }

    enum ResponseError: Error {
        case requestUnsuccessful
        case unexpectedResponseStructure
    }

    class ReposTableViewController: UITableViewController {
        internal var session = URLSession.shared
        //...
    }
```

You may notice something a little different about the preceding functions since we're now making full use of Swift's `Codable` protocol. With `Codable`, we can map the JSON response from our API straight to our struct models, without the need to convert this into a dictionary and then iterate each key-value pair to a property.

7. Next, in our table view, each row of the table view will display the name of one of the repositories that we retrieve from the GitHub API. We need a place to store the repositories that we retrieve from the API:

```
    class ReposTableViewController: UITableViewController {
        internal var session = URLSession.shared
        internal var repos = [Repo]()
        //...
    }
```

The `repos` array has an initially empty array value, but we will use this property to hold the fetched results from the API.

We don't need to fetch the repository data right now. So, instead, we'll learn how to provide data to be used in the table view. Let's get started.

8. Let's create a couple of fake repositories so that we can temporarily populate our table view:

```
    class ReposTableViewController: UITableViewController {
        let session = URLSession.shared
        var repos = [Repo]()

        override func viewDidLoad() {
            super.viewDidLoad()
```

```
        let repo1 = Repo(name: "Test repo 1", url: URL(string:
"http://example.com/repo1")!)
        let repo2 = Repo(name: "Test repo 2", url: URL(string:
"http://example.com/repo2")!)
        repos.append(contentsOf: [repo1, repo2])
    }
    //...
}
```

The information in a table view is populated from the table view's data source, which can be any object that conforms to the UITableViewDataSource protocol.

When the table view is displayed and the user interacts with it, the table view will ask the data source for the information it needs to populate the table view. For simple table view implementations, it is often the view controller that controls the table view that acts as the data source. In fact, when you create a subclass of UITableViewController, as we have, the view controller already conforms to UITableViewDataSource and is assigned as the table view's data source.

9. Some of the methods defined in UITableViewDataSource were created as part of the UITableViewController template; the three we will take a look at are as follows:

```
override func numberOfSections(in tableView: UITableView) -> Int
{
    // #warning Incomplete implementation, return the number of
sections
    return 0
}


override func tableView(_ tableView: UITableView,
numberOfRowsInSection section: Int) -> Int {
    // #warning Incomplete implementation, return the number of
rows
    return 0
}


/*
override func tableView(_ tableView: UITableView, cellForRowAt
indexPath: IndexPath) -> UITableViewCell {
    let cell = tableView.dequeueReusableCell(
withIdentifier: "RepoCell", for: indexPath)
    // Configure the cell...
    return cell
}

*/
```

Data in a table view can be divided into sections, and information is presented in rows within those sections; information is referenced through `IndexPath`, which consists of a section integer value and a row integer value.

10. The first thing that the data source methods ask us to provide is the number of sections that the table view will have. Our app will only be displaying a simple list of repositories, and as such, we only need one section, so we will return `1` from this method:

```
override func numberOfSections(in tableView: UITableView) -> Int
{
    return 1
}
```

11. The next thing we have to provide is the number of rows the table view should have for a given section. If we had multiple sections, we could examine the provided section index and return the right number of rows, but since we only have one section, we can return the same number in all scenarios.

We are displaying all the repositories we have retrieved, so the number of rows is simply the number of repositories in the `repos` array:

```
override func tableView(_ tableView: UITableView,
numberOfRowsInSection section: Int) -> Int {
    return repos.count
}
```

> **Tip**
> Notice that in the preceding two functions, we no longer use the return keyword. This is because, starting with *Swift 5.1*, you can now use implicit returns in functions. As long as your function doesn't carry ambiguity about what should and should not be returned, the compiler can work this out for you. This allows for more streamlined syntax.

Now that we have told the table view how many pieces of information to display, we must be able to display that information. A table view displays information in a type of view called `UITableViewCell`, and this cell is what we have to provide next.

For each index path within the section and row bounds that we have provided, we will be asked to provide a cell that will be displayed by the table view. A table view can be very large in size as it may need to represent a large amount of data. However, there are only a handful of cells that can be displayed to the user at any one time. This is because only a portion of the table view can be visible at any one time:

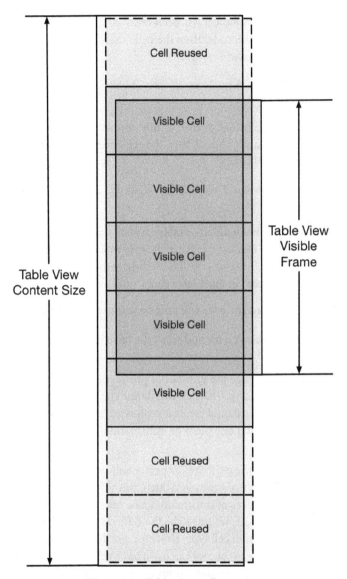

Figure 7.9 – Table view cell overview

In order to be efficient and prevent your app from slowing down as the user scrolls, the table view can reuse cells that have already been created but have since moved off-screen. Implementing cell reuse happens in two stages:

- Registering the cell's type with the table view with a reuse identifier.
- Dequeuing a cell for a given reuse identifier. This will return a cell that has moved off-screen or create a new cell if none are available for reuse.

How a cell is registered will depend on how it has been created. If the cell has been created and its subviews have also been laid out in the code, then the cell's class is registered with the table view through this method on `UITableView`:

```
func register(_ cellClass: AnyClass?, forCellReuseIdentifier
identifier: String)
```

If the cell has been laid out in `.xib` (usually called a "nib" for historical reasons), which is a visual layout file for views that's similar to a storyboard, then the cell's `nib` is registered with the table view through this method on `UITableView`:

```
func register(_ nib: UINib?, forCellReuseIdentifier identifier:
String)
```

Lastly, cells can be defined and laid out within the table view in a storyboard. One advantage of this approach is that there is no need to manually register the cell, as with the previous two approaches; registering with the table view is free. However, one disadvantage of this approach is that the cell layout is tied to the table view, so it can't be reused in other table views, unlike the previous two implementations.

Let's lay out our cell in the storyboard since we will only be using it with one table view:

1. Switch to our `Main.storyboard` file and select the table view in our `ReposTableView-Controller`.

2. In the attributes inspector, change the number of prototype cells to `1`; this will add a cell to the table view in the main window. This cell will define the layout of all the cells that will be displayed in our table view. You should create a prototype cell for each type of cell layout you will need; we are only displaying one piece of information in our table view, so all our cells will be of the same type.

3. Select a cell in the storyboard. The attributes inspector will switch to showing the attributes for the cell. The cell style will be set to custom, and often, this will be what you want it to be. When you are displaying multiple pieces of information in a cell, you will usually want to create a subclass of `UITableViewCell`, set this to be the cell's class in the class inspector, and then lay out `subviews` in this custom cell type. However, for this example, we just want to show the name of the repository. Due to this, we can use a basic cell style that just has one text label, without a custom subclass, so choose **Basic** from the **Style** dropdown.

4. We need to set the reuse identifier that we will use to dequeue the cell later, so type an appropriate string, such as `RepoCell`, into the reuse **Identifier** box of the attributes inspector:

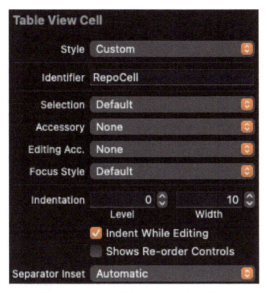

Figure 7.10 – Table view cell identifier

5. Now that we have a cell that is registered for reuse with the table view, we can go back to our view controller and complete our conformance with UITableViewDataSource.

6. Our ReposTableViewController contains some commented code that was created as part of the template:

```
/*
override func tableView(_ tableView: UITableView, cellForRowAt
indexPath: IndexPath) -> UITableViewCell {
    let cell = tableView.dequeueReusableCell(
withIdentifier: "RepoCell", for: indexPath)
    // Configure the cell...
    return cell
}
*/
```

7. At this point, you can remove the /* */ comment signifiers as we are ready to implement this method.

    This data source method will be called every time the table view needs to place a cell on-screen; this will happen the first time the table is displayed as it needs cells to fill the visible part of the table view. It will also be called when the user scrolls the table view in a way that will reveal a new cell so that it becomes visible.

8. Regarding the method's definition, we can see that we are provided with the table view in question and the index path of the cell that is needed, and we are expected to return UITableViewCell.

The code provided by the template actually does most of the work for us; we just need to provide the reuse identifier that we set in the storyboard and set the title label of the cell so that we have the name of the correct repository:

```
override func tableView(_ tableView: UITableView, cellForRowAt
indexPath: IndexPath) -> UITableViewCell {
    let cell = tableView.dequeueReusableCell(
withIdentifier: "RepoCell", for: indexPath)
    // Configure the cell...
    let repo = repos[indexPath.row]
    cell.textLabel?.text = repo.name
    return cell
}
```

The cell's textLabel property is optional because it only exists when the cell's style is not custom.

9.   Since we've now provided everything the table view needs to display our repository information, let's click on **Build and Run** and take a look:

Figure 7.11 – Our app's first run

Great! Now that we have our two test repositories displayed in our table view, let's replace our test data with real repositories from the GitHub API.

We added our `fetchRepos` method earlier, so all we need to do is call this method, set the results to our `repos` property, and tell our table view that it needs to reload since the data has changed:

```swift
class ReposTableViewController: UITableViewController {
    internal var session = URLSession.shared
    internal var repos = [Repo]()

    override func viewDidLoad() {
        super.viewDidLoad()
        title = "Repos"
        fetchRepos(forUsername: "SwiftProgrammingCookbook"){ [weak
self] result in
            switch result {
            case .success(let repos):
                self?.repos = repos
            case .failure(let error):
                self?.repos = []
                print("There was an error: \(error)")
            }
            self?.tableView.reloadData()
        }
    }
    //...
}
```

As we did in previous recipes, we fetched the repositories from the GitHub API and received a result in `enum` informing us of whether this was a success or a failure. If it was successful, we store the resulting `repository` array in our `repos` property. Once we have handled the response, we call the `reloadData` method on `UITableView`, which instructs the table view to requery its source for cells to display.

We also provided a weak reference to `self` in our closure's capture list to prevent a retain cycle. You can find out more about why this is important in the *Passing around functionality with closures* recipe of *Chapter 1, Swift Fundamentals*.

At this point, there is an important consideration that needs to be addressed. The iOS platform is a multithreaded environment, which means that it can do more than one thing at once. This is critical to being able to maintain a responsive UI, while also being able to process data and perform long-running tasks. The iOS system uses queues to manage this work and reserves the "main" queue for any work involving the UI. Therefore, any time you need to interact with the UI, it is important that this work is done from the main queue.

Our `fetchRepos` method presents a situation where this might not be true. The `fetchRepos` method performs networking, and we provide closure to `URLSession` as part of creating a `URLSessionDataTask` instance, but there is no guarantee that this closure will be executed on the main thread. Therefore, when we receive a response from `fetchRepos`, we need to *dispatch* the work of handling that response to the main queue to ensure that our updates to the UI happen on the main queue. We can do this using the `Dispatch` framework, so we need to import that at the top of the file:

```
class ReposTableViewController: UITableViewController {
    let session = URLSession.shared
    var repos = [Repo]()
    override func viewDidLoad() {
        super.viewDidLoad()
        title = "Repos"
        fetchRepos(forUsername: "SwiftProgrammingCookbook"){ [weak
self] result in
            DispatchQueue.main.async {
                switch result {
                case .success(let repos):
                    self?.repos = repos
                  case .failure(let error):
                    self?.repos = []
                    print("There was an error: \(error)")
                }
                self?.tableView.reloadData()
            }
        }
    }
}
```

We discussed multithreading and the `Dispatch` framework in greater depth in *Chapter 6, Understanding Concurrency in Swift*. So, let's jump right in:

1. Click on **Build and Run**. After a few seconds, the table view will be filled with the names of various repositories from the GitHub API.

    Now that we have repositories being displayed to the user, the next piece of functionality we'll implement for our app is the ability to tap on a cell and have it display the repository's GitHub page in a WebView.

    Actions triggered by the table view, such as when a user taps on a cell, are provided to the table view's delegate, which can be anything that conforms to `UITableViewDelegate`. As was the case with the table view's data source, our `ReposTableViewController` already conforms to `UITableViewDelegate` because it is a subclass of `UITableViewController`.

2. If you take a look at the documentation for the `UITableViewDelegate` protocol, you will see a lot of optional methods; this documentation can be found at `https://developer.apple.com/reference/uikit/uitableviewdelegate`. The one that's relevant for our purposes is as follows:

```
func tableView(_ tableView: UITableView, didSelectRowAt
indexPath: IndexPath)
```

3. This will be called on the table view's delegate whenever a cell is selected by the user, so let's implement this in our view controller:

```
override func tableView(_ tableView: UITableView, didSelectRowAt
indexPath: IndexPath) {
    let repo = repos[indexPath.row]
    let repoURL = repo.url
    // TODO: Present the repo's URL in a webview
}
```

4. For the functionality it provides, we will use `SFSafariViewController`, passing it the repository's URL. Then, we will pass that view controller to the `show` method, which will present the view controller in the most appropriate way:

```
override func tableView(_ tableView: UITableView, didSelectRowAt
indexPath: IndexPath) {
    let repo = repos[indexPath.row]
    guard let repoURL = repo.url else {
        return
    }
    let webViewController = SFSafariViewController(
url: repoURL)
    show(webViewController, sender: nil)
}
```

5. Make sure you import `SafariServices` at the top of the file.

6. Click on **Build and Run**, and once the repositories are loaded, tap on one of the cells. A new view controller will be pushed onto the screen, and the relevant repository web page will load.

Congratulations – you've just built your first app and it looks great!

How it works...

Currently, our app fetches repositories from a specific, hardcoded GitHub username. It would be great if, rather than hardcoding the username, the user of the app could enter the GitHub username that the repositories will be retrieved for. So, let's add this functionality:

1. First, we need a way for the user to enter their GitHub username; the most appropriate way to allow a user to enter a small amount of text is through the use of `UITextField`.

2. In the main storyboard, find **Text Field** in the object library, drag it over to the main window, and drop it onto the navigation bar of our `ReposTableViewController`. Now, you need to increase the width of **Text Field**. For now, just hardcode this to around `300px` by highlighting **Text Field** and selecting the **Size Inspector** option:

Figure 7.12 – Adding a UITextField instance

Like a table view, `UITextField` communicates user events through a delegate, which needs to conform to `UITextFieldDelegate`.

3. Let's switch back to `ReposTableViewController` and add conformance to `UITextFieldDelegate`; it is a common practice to add protocol conformance to an extension, so add the following at the bottom of `ReposTableViewController`:

    ```
    extension ReposTableViewController: UITextFieldDelegate {


    }
    ```

4. With this conformance in place, we need to set our view controller to be the delegate of `UITextField`. Head back over to the main storyboard, select the text field, and then open **Connections Inspector**. You will see that the text field has an outlet for its delegate property. Now, click, hold, and drag from the circle next to our delegate over to the symbol representing our `Repos Table View Controller`:

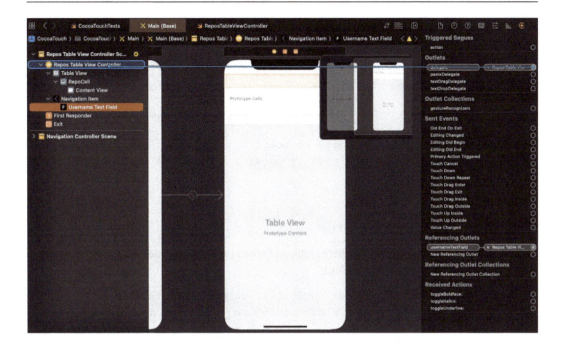

Figure 7.13 – UITextField with IBOutlet

The **delegate** outlet should now have a value:

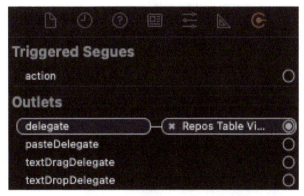

Figure 7.14 – UITextField delegate outlet

By taking a look at the documentation for `UITextFieldDelegate`, we can see that the `textFieldShouldReturn` method is called when the user presses the *Return* button on their keyboard after entering text, so this is the method we will implement.

5. Let's switch back to our `ReposTableViewController` and implement this method in our extension:

```
extension ReposTableViewController: UITextFieldDelegate {
    public func textFieldShouldReturn(_ textField: UITextField)
-> Bool {
        // TODO: Fetch repositories from username entered into
text field
        // TODO: Dismiss keyboard
        // Returning true as we want the system to have the
default behaviour
        return true
    }
}
```

6. Since repositories will be fetched here instead of when the view is loaded, let's move the code from `viewDidLoad` to this method:

```
extension ReposTableViewController: UITextFieldDelegate {
    public func textFieldShouldReturn(_ textField: UITextField)
-> Bool {
        // If no username, clear the data
        guard let enteredUsername = textField.text else {
            repos.removeAll()
            tableView.reloadData()
            return true
        }
        // Fetch repositories from username entered into text
field
        fetchRepos(forUsername: enteredUsername) { [weak self]
result in
            switch result {
            case .success(let repos):
                self?.repos = repos
            case .failure(let error):
                self?.repos = []
                print("There was an error: \(error)")
            }
            DispatchQueue.main.async {
                self?.tableView.reloadData()
            }
        }

        // TODO: Dismiss keyboard
        // Returning true as we want the system to have the
default behaviour
```

```
            return true
        }
    }
```

Cocoa Touch implements the programming design pattern **MVC**, which stands for **Model View Controller**; it is a way of structuring your code to keep its elements reusable, with well-defined responsibilities. In the MVC pattern, all code related to displaying information falls broadly into three areas of responsibility:

- **Model** objects hold the data that will eventually be displayed on the screen; this might be data that was retrieved from the network or device, or that was generated when the app was running. These objects may be used in multiple places in the app, where different view representations of the same data may be required.

- **View** objects represent the UI elements that are displayed on the screen; these may just display information that they are provided, or capture input from the user. View objects can be used in multiple places where the same visual element is needed, even if it is showing different data.

- **Controller** objects act as bridges between the models and the views; they are responsible for obtaining the relevant model objects and for providing the data to be displayed to the right view objects at the right time. Controller objects are also responsible for handling user input from the views and updating the model objects as needed:

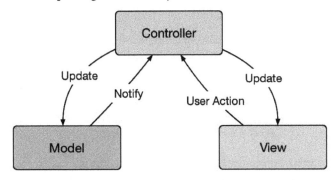

Figure 7.15 – MVC overview

With regard to displaying web content, our app provides us with a number of options:

- `WKWebView`, provided by the WebKit framework, is a view that uses the latest rendering and JavaScript engine for loading and displaying web content. While it is newer, it is less mature in some respects and has issues with caching content.

- `SFSafariViewController`, provided by the `SafariServices` framework, is a view controller that displays web content and also provides many of the features that are available in *Mobile Safari*, including sharing and adding to reading lists and bookmarks. It also provides a convenient button for opening the current site in Mobile Safari.

There's more...

The last thing we need to do is dismiss the keyboard. Cocoa Touch refers to the object that is currently receiving user events as the first responder. Currently, this is the text field.

It's the act of the text field becoming the first responder that caused the keyboard to appear on-screen. Therefore, to dismiss the keyboard, the text field just needs to resign its place as first responder:

```
extension ReposTableViewController: UITextFieldDelegate {
    public func textFieldShouldReturn(_ textField: UITextField) ->
Bool {
        textField.resignFirstResponder()
        return true
    }
}
```

Now, click on **Build and Run**. At this point, you can enter any GitHub account name in the text field to retrieve a list of its public repositories. Note that if your Xcode simulator doesn't have `soft keyboard` enabled, you can just press *Enter* on your physical keyboard to search for the repo.

See also

For more information regarding what was covered in this recipe, please refer to the following links:

- Apple documentation for GCD: `https://developer.apple.com/documentation/dispatch`

- Apple documentation for UIKit: `https://developer.apple.com/documentation/uikit`

Unit and integration testing with XCTest

Testing plays a massive role in the software development life cycle. Primarily, there should be a lot of focus on physical user testing – putting your piece of code in the hands of those who use it day in and day out. This should be one of our main focuses, but what about testing what we, as software developers, do? How do we test and check the integrity of our code base?

This is where unit testing and integration testing come in. In this recipe, we'll cook up a unit and integration test for our previously written Cocoa Touch app. This will be written entirely in Swift, using the Xcode IDE.

Getting ready

Back in our existing `CocoaTouch` project, in the **File** inspector, look for a folder called `CocoaTouchTest`. Expand this and select the `CocoaTouchTests.swift` file. Inside this file, you'll notice a `CocoaTouchTests` class, which, in turn, inherits from the `XCTestCase` class. `XCTestCase` offers a suite of functions that we can use when writing our unit tests.

So, what exactly is a **unit test**? Well, it's a test (or in our case, just a function) that checks that another function is doing what it's supposed to be doing. Writing tests or functions using XCTestCase allows us to not only use the previously mentioned suite of helpers but also allows Xcode to visualize and report on metrics such as test coverage.

With that, let's get stuck into cooking up our first unit test! In the CocoaTouchTests.swift file, you'll see some override functions that have already been generated by Xcode. Just ignore these for now; we'll work on them as and when we need to.

How to do it...

Let's start by creating the following function:

```
func testThatRepoIsNotNil() {
    XCTAssertNotNil(viewControllerUnderTest?.repos)
}
```

So, let's go through this one bit at a time. We'll start with the testThatRepoIsNotNil function. The common practice when naming a unit test is for the name to be as descriptive as possible. Depending on your coding standard, you can choose to either camel case these or snake case them (I much prefer camel case), but when writing tests with Xcode, you always have to prefix these tests with test.

So, what are we testing? Here, we're checking that our repos array is not nil.

Looking back at our ReposTableViewController, you'll remember that we instantiated our repo model where the variable was declared, so this is a great test to start with. Let's say someone tries to change this to an optional, like this:

```
internal var repos: [Repo]?
```

If this happens, the code in our CocoaTouch app will compile, but the test will fail.

Let's take another look at our test. Note that the function we're calling to check our repo model is viewControllerUnderTest. This is how we access our RepoTableViewContoller. We can achieve this by adding the following class-level variable to our file:

```
var viewControllerUnderTest: ReposTableViewController?
```

Now, we need to instantiate this. Add the following override method from XCTestCase to your class:

```
override func setUp() {
    viewControllerUnderTest = ReposTableViewController()
}
```

When running your unit test for this particular class, `setUp()` will run before any of your test cases run, allowing you to prep anything you may need, such as instantiating a class. Once the tests are complete and you want to free anything up or close anything down, you can simply do this with the `tearDown()` function.

This was and is a very basic test, but the main purpose here was not necessarily to look at testing practice, but at how we'd do it in Swift. However, before we go any further, let's take a look at the `Assert` options that are available to us.

Previously, we used `XCTAssertNotNil`, which worked perfectly for our scenario. However, the following options are also available:

- `XCTAssert`

- `XCTAssertEqual`

- `XCTAssertTrue`

- `XCTAssertGreaterThan`

- `XCTAssertGreaterThanOrEqual`

- `XCTAssertLessThan`

- `XCTAssertLessThanOrEqual`

- `XCTAssertNil`

These are just a handful of the common ones and they are pretty self-explanatory – an added bonus is that each one has an optional `message` parameter, which allows you to add a custom string. This allows you to be more specific about the assertion that took place (this is ideal for reporting in a CI/CD world).

Now that we understand the basics of how to write tests in Swift, we need to learn how to run them. There are two ways we can achieve this:

- First, we can run all the tests in our class in one go. We can do this by simply clicking on the diamond to the left of the class's declaration:

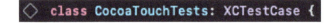

Figure 7.16 – Class test case

- If we want to run tests individually, then we can simply select the icon next to our individual test case, like this:

Figure 7.17 – Method test case

If everything goes to plan and our tests pass, we'll see the icon turn green:

Figure 7.18 – Method passed test case

However, if one or more of the tests in our class fail, we'll see the icon turn red:

Figure 7.19 – Method failed test case

Alternatively, the *CMD + U* keyboard shortcut will also get Xcode to run any tests associated with the main project. Remember, only functions that start with `test` will be treated as a test case (excluding the class name), so feel free to add a private function in your test case should you need to.

Next, let's take a look at how we would test networking logic in Swift, using mock data to help us out:

1. We'll start by creating the following test function:

   ```
   func testThatFetchRepoParsesSuccessfulData() { }
   ```

2. Let's start by figuring out how we are going to call this. Once again, we'll take advantage of our `viewControllerUnderTest` variable:

   ```
   func testThatFetchRepoParsesSuccessfulData() {
           viewControllerUnderTest?.fetchRepos(forUsername: "",
   completionHandler: { (response) in
               print("\(response)")
           })
   }
   ```

   This works as expected, but unfortunately, it's not that simple; this will simply call the API just like our app would. If we were to add any `XCAssert` instances inside our code, they wouldn't be executed as our test and function would have finished and been torn down before the API had a chance to respond.

3. To do this, we need to mock some objects in our `viewControllerUnderTest`, starting with `URLSession` and `URLSessionDataTask`. So, why do we need to mock these two? Let's start by taking a look at where we use them in our `CocoaTouch` app:

   ```
   let task = session.dataTask(with: request) { (data, response,
   error) in
   ```

   Here, we are using `URLSession` and one of its functions, `URLSessionDataTask`, by mocking `URLSession`. We're creating our own local session here that we can then use to call our `MockURLSessionDataTask`. So, the real question here is, what is our

MockURLSessionDataTask doing? We're using this to pass in some mock data – data that we should expect from the API – and then running this through our logic. This guarantees the integrity of our tests every time!

4. We could create the following input in our own files, but for now, we'll just append them to the bottom of our CocoaTouchTests.swift file. First, let's look at our MockURLSession:

```swift
class MockURLSession: URLSession {
    override func dataTask(with request: URLRequest,
completionHandler: @escaping (Data?, URLResponse?, Error?) ->
Void) -> URLSessionDataTask {
        return MockURLSessionDataTask(
completionHandler: completionHandler, request: request)
    }
}
```

The preceding function is pretty self-explanatory – we simply override the dataTask() function with the following MockURLSessionDataTask:

```swift
class MockURLSessionDataTask: URLSessionDataTask {
    var completionHandler: (Data?, URLResponse?, Error?) ->
Void
    var request: URLRequest
    init(completionHandler: @escaping (Data?, URLResponse?,
Error?) -> Void, request: URLRequest) {
        self.completionHandler = completionHandler
        self.request = request
        super.init()
    }

    var calledResume = false
    override func resume() {
        calledResume = true
    }
}
```

At first glance, this looks a little complex, but all we are really doing here is adding our own completionHandler. This will allow it to be called synchronously from our test (stopping our test from running away from us).

Let's put this all together and head back over to our new test:

1. Let's start by setting our MockURLSession for our viewControllerUnderTest. This is nice and simple. Now, line by line, add the following:

```swift
func testThatFetchRepoParsesSuccessfulData() {
    viewControllerUnderTest?.session = MockURLSession()
    // ...
}
```

2. Let's start by adding in our main `responseObject`. This is what we are going to perform our `XCAssert` instances against. Declare this as an optional variable:

```
var responseObject: FetchReposResult?
```

3. Now, we can call our function, much like we tried to earlier in this section. However, this time, we'll assign the result to a variable and cast this as a `MockURLSessionDataTask` instance:

```
let result = viewControllerUnderTest?.fetchRepos(
forUsername: "", completionHandler: { (response) in
    responseObject = response
}) as? MockURLSessionDataTask
```

4. Remember that we can pass in anything we want to the `userName` variable as we're not going to be calling the API. Now, let's fire the completion handler we created and force through our `mockData`:

```
result?.completionHandler(mockData, nil, nil)
```

I've highlighted the `mockData` variable in the preceding code as we'll need to add this to the JSON response we want to test against. You can get this by simply visiting the GitHub URL and copying this into a new, empty file in the project. I did this for my username and created a file called `mock_Data.json`:

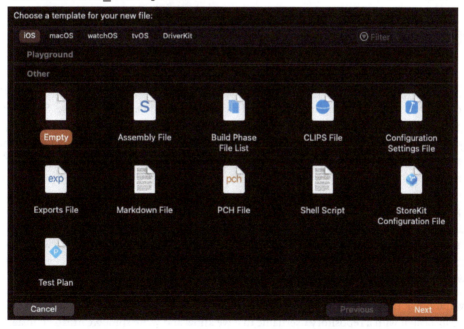

Figure 7.20 – Adding an empty file

Remember to select the `CocoaTouch` target when you're saving the file to disk; otherwise, the following steps won't work.

5. Now, create a computed property in our `Test` class that simply reads in the file and spits out the `Data()` object:

```
var mockData: Data {
      if let path = Bundle.main.path(forResource: "mock_Data",
ofType: "json"),
      let contents = FileManager.default.contents(atPath:
path) {
            return contents
      }
      return Data()
}
```

6. At this point, we can successfully pass our mock data through our `fetchRepos` function without the need to call the API. All we need to do now is write some asserts:

```
switch responseObject {
case .success(let repos):
      // Our test data had 3 repos, lets check that parsed okay
      XCTAssertEqual(repos.count, 9)
      // We know the first repo has a specific name... let's
check that
      XCTAssertEqual(repos.first?.name, "aerogear-ios-oauth2")
      default:
            // Anything other than success is a failure
            XCTFail()
}
```

What you test for here is really up to you – it's all based on the test cases you choose. Sometimes, thinking about what to test when you've already written a function can be a hard task. As a developer, it is easy for you to get too close to the project. This is where **test-driven development** (**TDD**) comes in, a methodology for writing tests before writing any code at all. Let's take a look at this and what we can achieve with it.

How it works...

Testing networking logic can be troublesome. I find that questions always arise, such as, what should you test? What exactly is being tested? However, if you can get your head around these questions, then you're well on your way to understanding the core fundamentals of unit testing.

Let's try and break this down. The logic we want to test is our `fetchRepos()` function. This is easy – we just call it with a repository username that we know and write some `XCAssert` instances against the list of repositories that come back.

While that will work for now, what happens when the user removes a repository? Your test will fail. This isn't good because your logic is not actually flawed – it's just the data that is wrong, much like if the API decided to return some malformed JSON due to an internal server error. This isn't your code's fault – it's the API's fault, and it's the API's job to make sure that it works.

All you want to do is check that if the server gives you a specific response, with a specific piece of data, your logic does what it says it should do. So, how can we guarantee the integrity of the data coming back from an API? We can't – that's why we mock up the data ourselves and, in turn, don't actually call the service at all.

There's more…

TDD is a methodology that includes writing your unit test first, before actually writing your desired function. Some believe this is the only way to write code, while others say it should be used only when necessary. For the record, I prefer the latter, but we're not here to get into the theory – we're here to learn how to achieve this in Swift using `XCTest`.

Going back to our `CocoaTouch` app, let's say we'd like to write a function that validates `UITextField` for whitespaces. Perform the following steps to achieve this:

1. We'll start by writing out a stub function, which will look something like this:

```
func isUserInputValid(withText text: String) -> Bool {
    return false
}
```

Normally, here, I would litter my function with comments about what I'd like to achieve, but for TDD, we're going to do this the other way around.

2. Back inside the `CocoaTouchTests.swift` file, add the following test:

```
func testThatTextInputValidatesWithSingleWhitespace() {
}
```

Again, taking the name of our test as a literal description, we're going to check that our function correctly detects whitespaces in the middle of a `String()` variable.

3. So, let's write a test against the current function:

```
func testThatTextInputValidatesWithSingleWhitespaces() {
    let result = viewControllerUnderTest?.
isUserInputValid(withText: "multiple white spaces")
    XCTAssertFalse(result!)
}
```

4. With that, we're happy we've asserted everything we set out to do in our test cases. Now, we can go ahead and run our test.

As expected, our test will fail, which is obvious for two reasons. First, we didn't really write up our function, and second, we hardcoded the return type as false.

We actually hardcoded the return type as false on purpose, because the TDD methodology is done in three stages:

1. **Fail test**: *Done, we did that.*

2. **Pass test**: *Can be as messy as you like.*

3. **Refactor code**: *We can do this with the utmost confidence.*

The idea is to write your unit test to cover all the scenarios and asserts that may be required for that test case and make it fail (like we did).

With the foundations set up, we can now confidently move over to our function and code away, safe in the knowledge that we'll be able to run our test to check whether our function is broken or not:

```
func isUserInputValid(withText text: String) -> Bool {
    return !text.contains(" ")
}
```

This is nothing special, but for the purposes of this section, it doesn't need to be. TDD with Swift doesn't have to be daunting. After all, it's just a methodology that works perfectly well with XCTest.

See also

You can find more information about unit testing at `https://developer.apple.com/documentation/xctest`.

UI testing with XCUITest

UI testing has been around for a while. In theory, it's done every day by anyone who is using, testing, or checking an app, but in terms of automation, it's had its fair share of critics.

However, with Swift and XCTest, UI testing has never been easier, and the beauty of this approach is that it also has an amazing hidden benefit.

Getting ready

Unlike unit testing, when we are testing against a function, piece of logic, or algorithm, UI tests are exactly what they say on the tin. They are a way for us to test the UI and UX of the app – things that might not necessarily have been generated programmatically.

Head on over to the `CocoaTouchUITests.swift` file that was automatically generated when we created our project. Again, much like the unit tests, you'll notice some placeholder functions in there. We'll start by taking a look at one called `testExample()`.

How to do it...

With what we mentioned in the *Getting ready* section in mind, let's take a look at our app and see what we would like to test. The first thing that comes to my mind is the search bar:

Figure 7.21 – Search bar selected

Now that we've made it mandatory to populate this in order for our app to work, we want to make sure it is here all the time, so let's write a UI test for this:

```
func testExample() throws {
    // UI tests must launch the application that they test.
    let     app = XCUIApplication()
    app.launch()
}
```

As the comment correctly states, in order for the tests to be successful, the app needs to be launched, which is taken care of by the `launch()` function. However, once our app has been launched, how do we tell it to check for a `UITextField` instance and, more importantly, a specific one (in the future, we could have multiple instances on our screen)?

To do this, we must start with the basics:

1. I've edited the name of the function to make it more applicable to what we are testing here. As you can see from the following highlighted code, we've told our test to select the `textFields` element and tap it:

```
func testThatUsernameSearchBarIsAvailable() throws {
    let app = XCUIApplication()
    app.launch()
    app.textFields.element.tap()
}
```

2. Go ahead and run the test by clicking on the diamond to the left to watch your app come to life in the simulator. If you're quick enough, you'll see the cursor enter the text box just before the app closes.

Great work! The test passed, which means you've written your first UI test.

Regarding our previous test, we weren't specific about the element being identified. For now, this is okay, but building a much bigger and more complex app may require that you test certain aspects of specific elements. Let's take a look at how we could achieve this:

1. One way would be to set an accessibility identifier for our `UITextField` – a specific identifier that's required for accessibility purposes that, in turn, will allow our UI test to identify the control we want to test.

2. Back over in our `RepoTableViewController.swift` file, create an `IBOutlet` instance for the `UITextField` object in question and add the following code, remembering to hook up the outlet to your `ViewController`:

```
@IBOutlet weak var usernameTextField: UITextField! {
    didSet {
        usernameTextField.accessibilityIdentifier = "input.
textfield.username"
    }
}
```

3. With that in place, comment out or replace our generic `UTextField` tap test with the following:

```
app.textFields.element(matching: .textField, identifier: "input.
textfield.username").tap()
```

4. Now, run your test and watch it pass. Great stuff!

    Notice that we are identifying a `textField` variable and then matching a control type from `textField`. This approach will work wonders when we're testing for nested components in specific views of your app. For example, you might want to search and match for a specific `UIButton` instance that you know is embedded within a specific `UIScrollView` instance:

    ```
    app.scrollViews.element(matching: .button, identifier: "action.
    button.stopscrolling").tap()
    ```

5. With that done, let's take our test a little further. Notice the `.tap()` function we're calling at the end of our element identification. There are plenty more options to choose from, but we'll start by creating a reference of element by capturing it, in its own variable:

    ```
    let textField = app.textFields.element(matching: .textField,
    identifier: "input.textfield.username")
    ```

6. Notice that we've removed the `.tap()` function. Now, we can simply call this and any other available function via our `textField` variable:

    ```
    textField.tap()
    textField.typeText("MrChrisBarker")
    ```

7. Run this to see it in action for yourself. Now, what if we go a little further? Add the following line and run the code once more:

    ```
    app.keyboards.buttons["return"].tap()
    ```

    Hopefully, at this point, you can see where we are going. One thing to bear in mind is that, since we are not mocking up data here, we're making a live, asynchronous API call, which, depending on your connection speed or the API, could vary from test to test.

8. To check the results, we need our UI test to wait for a specific element to come into view. By design, we know that we are expecting a `UITableView` instance with populated cells, so let's write our test based on that:

    ```
    let tableView = app.tables.staticTexts["XcodeValidateJson"]
    XCTAssertTrue(tableView.waitForExistence(timeout: 5))
    ```

    Line *1* of the preceding code should now be all but familiar to us – we're building an element based on cells within a `UITableView` instance (we're not being specific at this time) to look for a specific cell with a label of `XcodeValidateJson`.

    Then, we're using an `XCAssert` instance against this element. Allow for a timeout of five seconds for this to appear. If it appears beforehand, the test will pass; if not, it will fail.

There's more...

So far, we've seen how functions such as `.tap()` and `.typeText()` can be used when we're interacting with our app. However, these are not standard functions that we may use for `UIButton` and `UITextField`. When we're identifying our controls, the return type we get back is that of an `XCUIElement()` type.

There are more options available that we can use to enhance our UI tests, thus allowing for an intricate yet worthy automated test. Let's take a look at some of the additional options available to us:

- `tap()`
- `doubleTap()`
- `press()`
- `twoFingerTap()`
- `swipeUp()`
- `swipeDown()`
- `swipeLeft()`
- `swipeRight()`
- `pinch()`
- `rotate()`

Each comes with additional parameters that allow you to be specific and cover all the aspects of your user experience in the app (for example, `press()` has a duration parameter).

At the beginning of this section, I mentioned that UI tests come with a great additional benefit, and this is something we have seen already: accessibility. Accessibility is an important factor when building mobile apps, and Apple gives us the best possible tools to do this with Xcode and the Swift programming languages. However, from a theoretical perspective, if you take the time to build our accessibility into your app, you're indirectly making it much easier to build and shape the UI test around these identifiers – it will almost do a good 50% of the work for you – while including an amazing feature.

Alternatively, writing a good UI test can lead to improved accessibility in your app, making it really easy to have the one complement the other when building your app.

See also

You can find more information about XCUITest at `https://developer.apple.com/documentation/xctest/xcuielement`.

8

Building iOS Apps with SwiftUI

At the Apple **Worldwide Developers Conference** (**WWDC**) in 2019, Apple took a lot of us by surprise with the announcement of **SwiftUI**, a brand-new UI framework written from the ground up, entirely in Swift.

Making use of the declarative programming paradigm, SwiftUI not only offers a powerful way to programmatically create and design your UI but a functional and logical approach too.

Alongside many other announcements at WWDC 2019, Apple also announced its very own entry into the reactive programming stream with a new framework called **Combine**.

Combine replaces the traditional delegate pattern most of us will be accustomed to in iOS and macOS development. With SwiftUI's changes to the dynamics of how UI patterns are written programmatically, Combine is a welcome addition alongside the SwiftUI framework.

In this chapter, we'll take a tour of the inner workings of SwiftUI and how to build our very own app—alongside this, we'll integrate the power of Combine to give us a truly unique and reactive workflow. (We'll dive even deeper into Combine in *Chapter 9, Getting to Grips with Combine*.)

In this chapter, we will cover the following recipes:

- Declarative syntax
- Function builders, property wrappers, and opaque return types
- Building simple views in SwiftUI
- Combine and data flow in SwiftUI

Technical requirements

You can find the code files for this chapter on GitHub at `https://github.com/PacktPublishing/Swift-Cookbook-Third-Edition/tree/main/Chapter%208`.

Declarative syntax

With the introduction of SwiftUI comes a more modern coding paradigm called declarative syntax. In this section, we'll take a look at what exactly declarative syntax is and how it compares to the style of syntax we might be used to seeing already.

Getting ready

For this section, you'll need the latest version of Xcode available from the Mac App Store.

How to do it...

1. Open Xcode and select **File | New | Playground**, then select **Blank** in order to open a new playground canvas to work from.

2. Once that's open, add in the following syntax:

```
import PlaygroundSupport
import SwiftUI
```

We've seen the first `import` statement before, so it should be familiar. The next one is for SwiftUI—pretty self-explanatory as to why we need this.

3. Now, let's create a view in SwiftUI by adding the following code:

```
struct MyView: View {
    var body: some View {
        VStack {
            Text("Swift Cookbook")
        }
    }
}
```

All SwiftUI views are built in a struct that conforms to the `View` type—this then houses another struct, which looks a bit like a computed property, called `body`, which in turn conforms to some `View`. Inside this property—or "function builder," as it's known (which we'll touch on later in this chapter)—we have certain elements that start to make up our UI.

There is `VStack` or vertical stack, which will wrap all enclosing views "vertically" within itself. `VStack` is, again, a view.

Inside here, we have a `Text()` view where we set our text to be displayed.

4. If we add the following to our playground, we'll be able to see SwiftUI in action:

```
PlaygroundPage.current.setLiveView(MyView())
```

However, where does the declarative syntax come into all this? Well, it already has—you've written it, right there in your struct. Let's dig a little deeper into how declarative syntax works.

How it works...

In SwiftUI everything is made up of views—from the main container that is presented to the app's window, to text, a button, or even a toggle.

Thinking back to how UIKit works, this theory isn't too dissimilar—most objects are a subclass of `UIView()`.

The only fundamental difference is that with SwiftUI, the layout and construction of all this is much more visible; this is the declarative syntax coming into play. The best way to think about declarative syntax is in a functional and logical way.

I want to vertically align items in my view:

```
VStack {
    //...
}
```

I then want to add a `Text` box:

```
Text("Swift Cookbook")
```

Then, let's add a button:

```
Button(action: {
    print("Set Action Here...")
}, label: {
    Text("I'm going to perform an action")
})
```

Even the construction of the button is declarative itself: set an action; set a label. Everything is just... functional.

Another way to think of this would be similar to how we would work through a food recipe:

1. Chop onions.

2. Fry onions.

3. Add seasoning.

4. And so on...

With our more traditional style of programming (or imperative programming, as it's known), you might perform things a little differently and a little less logically:

1. Get seasoning.

2. Get onion.

3. Peel onion.

4. Chop onion.

5. Heat pan.

6. And so on...

While with declarative syntax all the preceding steps still need to exist to make it work, the framework that it is written in does a lot of the work for you—we just simply *tell it* what to do.

There's more...

Declarative syntax has been around for a while now; you may have used it before without even realizing it. Let's take a look at the following **Structured Query Language** (**SQL**) syntax:

```
SELECT column1, column2, ... FROM table_name WHERE condition;
```

Notice anything familiar? That's right: declarative syntax right there... give me `column1` and `column2` from a particular table where this condition is met.

Most recently, declarative syntax has been making its way into even more UI frameworks, such as Google's **Flutter** and, most recently, Android's new **Jetpack Compose**, both of which use a declarative syntax style to allow developers and designers to build a UI.

We've mentioned a few times already that declarative syntax gives us a much more functional and logical approach to programming. There are paradigms that sit beneath the declarative paradigm as a whole. SQL, for example, sits within **Domain-Specific Language** (**DSL**), along with HTML and other markup languages.

See also

- Android Jetpack Compose: `https://developer.android.com/jetpack/compose`
- Google's Flutter: `https://flutter.dev/`

Function builders, property wrappers, and opaque return types

SwiftUI certainly brings a lot to the table, especially as it's been built from the ground up using Swift at its core. This itself has a plethora of benefits, which include making use of some of the features we are about to cover in this section.

Getting ready

For this section, you'll need the latest version of Xcode available from the Mac App Store.

How to do it...

1. Continuing with our existing playground project, let's take another look at how things *stack up*. We'll start by taking another look at `VStack`:

```
VStack {
    Text("Swift Cookbook")
    Button(action: {
        print("Set Action Here...")
    }, label: {
        Text("I'm going to perform an action")
    })
}
```

Here is a block of code, which in SwiftUI terms is a view to be displayed. The view is a vertical stack—think `UITableView`, but at the same time don't think `UITableView`, as it's bad practice to try to compare SwiftUI to UIKit.

All the code sitting within `VStack` will be displayed vertically and then presented back to the main view, but where is the logic that adds our `Text()` and `Button()` views to `VStack`? There's no *item* or *row for index* (see, it's bad to compare this to `UITableView`); there's no `.add()` or `.append()` function that you would see when building an array. Everything just sits inside what are called function builders.

2. Let's add another in for good measure:

```
VStack {
    Text("Swift Cookbook")
    Button(action: {
        print("Set Action Here...")
    }, label: {
        Text("I'm going to perform an action")
    })
    HStack {
        Text("By Keith, Chris, & Danny")
        Image(systemName: "book")
    }
}
```

In the preceding code, we've added `HStack`, which (yep, you guessed it) gives us a horizontal stack of views—another function builder like before, this time housing a `Text()` and an `Image()` view.

Notice how we've added our `HStack` function builder inside the existing `VStack`? This is like we said before: our stacks are just views, so the top-level `VStack` just treats it like that and `HStack` does all the work of arranging its `Text()` and `Image()` views.

But what is being returned here? When building views in functions programmatically, you might expect to see the return keyword, with the return type specific to the object type being returned.

3. However, we can harness the power of opaque return types. Let's look back at the body of our SwiftUI view:

```
struct MyView: View {
    var body: some View {
        //...
    }
}
```

Notice the some View return type. This is an opaque return type and allows SwiftUI to return any type that conforms to the View protocol, such as Text, Button, Image, and so on. Without this, SwiftUI would not be as versatile in terms of allowing us to build up a view, and our view builder would simply not exist.

But the beauty of opaque return types is that they are not SwiftUI-specific; they are just a natural evolution of the Swift language, again demonstrating how much SwiftUI has been built from the core Swift programming language.

Another thing we see here is the omission of the return keyword. Our SwiftUI code can interpret a final return type to be passed back up the View hierarchy. But what about HStack and VStack? Well, as these are function builders, they are not returned, as such; it is more that they are added to the stack, which then, in turn, is passed back up.

However, there is always the possibility that you may need HStack sitting alongside VStack, like so:

```
VStack {
    //...
}
HStack {
    Text("I'm sitting underneath a HStack")
}
```

At this point, the compiler will need a little help. Unfortunately, we can't just add in a return keyword as we want both to be returned, so we could add these into another stack—but as we don't really need one, that would be unnecessary, so we simply wrap these in a Group() view instead:

```
Group {
    VStack {
        //...
    }
    HStack {
        Text("I'm sitting underneath a HStack")
    }
}
```

A Group() view is another view that can then be passed back up as some View to our body—nice!

4. We're certainly getting all the ingredients together in order to make a start with SwiftUI, but before we get stuck in, let's take a look at another feature introduced in SwiftUI, again from our ever-evolving Swift programming language.

Property wrappers in SwiftUI are one of the features that really help make it shine and are used for a wide variety of things. The main purpose that each one holds is to reduce the amount of maintenance required for your specific view. Let's take a look at some of the more common ones you might use:

- @State: @State allows SwiftUI to modify specific properties of specific views without the need to call a specific function to do so. For example, make the following changes to your code:

```swift
struct MyView: View {
    @State var count: Int = 0

    var body: some View {
        Group {
            VStack {
                Text("Swift Cookbook")

                Button(action: {
                    count += 1
                }, label: {
                    if count > 0 {
                        Text("Performed \(count) times")
                    } else {
                        Text("I'm going to perform an action")
                    }
                })

                HStack {
                    Text("By Keith, Chris, & Danny")
                    Image(systemName: "book")
                }
            }

            HStack {
                Text("I'm sitting underneath a HStack")
            }
        }
    }
}
```

We've added a variable called count and have given this the @State property wrapper, and we updated our button click to increase the integer by 1. Next, we added some logic based on the value of count.

By changing the value of count, we have bound our property that is being used within SwiftUI to the value and any changes that are made, thus invalidating the SwiftUI layout and rebuilding our view using the new value.

Go ahead—run this in the playground and try it out for yourself.

- @Binding: @Binding is another well-used property wrapper specifically used in conjunction with passing values to state properties that may live in another view. Let's take a look at how we might do this, starting by separating out some code and creating another SwiftUI view. We can do this just underneath the current MyView:

```
struct ResultView: View {
    @Binding var count: Int
    var body: some View {
        Text("Performed \(count) times")
    }
}
```

Here, we're simply creating a SwiftUI view that returns a Text view, but this is a great way to see how easy it is to separate out specific view logic that you might want to work on separately (or make reusable).

Notice here that we also used the count variable, although this time with the @Binding wrapper. This is because we won't be controlling the value of count from within this view; this will be done externally back in MyView:

```
struct MyView: View {
    @State var count: Int = 0
    var body: some View {
        Group {
            VStack {
                Text("Swift Cookbook")
                ResultView(count: $count)
                Button(action: {
                    count += 1
                }, label: {
                    Text("Perform Action")
                })
                HStack {
                    Text(«By Keith, Chris, & Danny»)
                    Image(systemName: «book»)
                }
            }
            HStack {
                Text("I'm sitting underneath a HStack")
            }
```

```
        }
    }
}
```

In the preceding highlighted code, notice we still have our `@State` variable, and our `Button` action is still updating this value on each press. We've also added in a new `ResultView`, passing in the `@State` variable and *binding* this to our variable in `ResultView`, thus forcing a change to that view every time `count` is updated. Go ahead and try it for yourself.

There's more...

We've covered some of the property wrappers that you are more than likely to be exposed to with SwiftUI from the outset, but there are plenty more where they came from—some of which we'll cover later on in this chapter, specifically when it comes to working with the Combine framework. However, here is a run-through of some of the others and what they have to offer:

- `@EnvironmentObject`: Think of this as a global object—sometimes you might want to keep track of certain things throughout your app that you might not necessarily need or feel the need to pass through to every view. However, it's important to know that `EnvironmentObject` isn't a single source of truth; it's data—it's merely referencing it from the source, and should the source change, `EnvironmentObject` will trigger a state change (which is what we want).

  Here's an example of how we could use this to create a class we want to observe, conforming to `ObservableObject`:

  ```
  class BookStatus: ObservableObject {
      @Published var released = true
      @Published var title = ""
      @Published var authors = [""]
  }
  ```

  Then, reference it from anywhere in our SwiftUI project, like this:

  ```
  @EnvironmentObject var bookStatus: BookStatus
  ```

- `@AppStorage`: Another great and certainly convenient property wrapper is `@AppStorage`, used as a way to access data stored within `UserDefaults`. We can incorporate this straight into our SwiftUI views, without the need for additional logic or functions. Let's take a look at how we would do this:

  ```
  @AppStorage("book.title")
  var title: String = "Book Title"
  ```

  Notice here that we have a default value should there be no data already persisted. If we want to write to this, we simply assign the property a value:

  ```
  title = "Swift Cookbook"
  ```

Go ahead and try this in your playground. If you get stuck, have a look at the GitHub resource to see how I did it.

Due to the architecture of SwiftUI, there is—and will be—an ever-growing list of available property wrappers. We'll cover some more later on in this chapter, but here are some others to be aware of:

- `@GestureState`: Tracks the current gesture that is being performed
- `@FetchRequest`: Performs a fetch for Core Data entities

See also

- States: `https://developer.apple.com/documentation/swiftui/state`
- Bindings: `https://developer.apple.com/documentation/swiftui/binding`

Building simple views in SwiftUI

We've covered some of the fundamentals of how SwiftUI is built up from the Swift programming language, but it's time now to get into how we build an actual app in SwiftUI.

In this section, we'll take everything we've learned so far and apply it, in order for us to build a list app similar to the one we created previously.

Getting ready

For this section, you'll need the latest version of Xcode from the Mac App Store.

How to do it...

1. Let's get going. First, we'll create a brand new project—in Xcode, click on **File | New | Project**. Then, select **Single View App** and make sure you've selected **SwiftUI** for the **Interface** style, just like I've done here:

Choose options for your new project:

Product Name:	TaskManager
Team:	Daniel Bolella (Personal Team)
Organization Identifier:	com.swift-cookbook
Bundle Identifier:	com.swift-cookbook.TaskManager
Interface:	SwiftUI
Language:	Swift

Use Core Data

Host in CloudKit

☑ Include Tests

Cancel   Previous   Next

Figure 8.1 – Creating a new project

2.   Click **Next** and select a location on your disk. Once that is done, the familiar sight of Xcode should appear; however, you may notice something new. On the right-hand side, you'll see the **Live Preview** screen - with the simulator. Go ahead and click **Resume**—you should see the following:

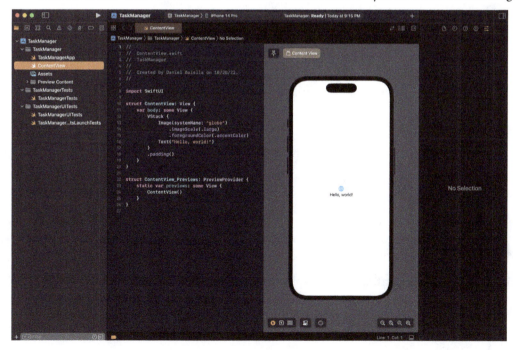

Figure 8.2 – Xcode and Live Window screen

Here, we've got a generated preview of our boilerplate SwiftUI code. Notice our `ContentView()` struct, just as we expect with its body. Now, look at the struct below it:

```
struct ContentView_Previews: PreviewProvider {
    static var previews: some View {
        ContentView()
    }
}
```

This is our `PreviewProvider` struct, allowing us to test out our SwiftUI views at design time, without the need to keep rerunning the simulator and rebuilding our application—neat!

Now, for our initial list, we're going to need some mock data.

3. Create the following struct (this can be in a new file if you want):

```
struct Task: Identifiable {
    var description: String
    var category: String
    var id = UUID()
}
```

Notice that we made our struct conform to `Identifiable` and be given a unique ID—this is required by SwiftUI for anything that we are going to iterate around.

4. Next, let's create a little helper function for some mock data:

```
struct MockHelper {
    static func getTasks() -> [Task] {
        var tasks = [Task]()
        tasks.append(Task(description: "Get Eggs", category:
"Shopping"))
        tasks.append(Task(description: "Get Milk", category:
"Shopping"))
        tasks.append(Task(description: "Go for a run",
category: "Health"))
        return tasks
    }
}
```

This mock data will come in handy in SwiftUI in more than one way, but we'll get to that shortly. Let's hook this up to our app.

5. Back over to our `ContentView`, replace the `Hello World` text view with the following:

```
List(MockHelper.getTasks()) { task in
    Text(task.description)
}
```

Notice something about our `List()` view? That's right—another function builder accepts an argument of an array of items, where the items conform to `Identifiable`. A variable is given back to us in the closure, representing each one of these items so that we can then use them inside our list builder as we see fit.

Here, we are just adding the description to a text view for now. If not already showing in the live preview, click **Resume** (sometimes this is needed in Xcode), and you should now see the following:

Figure 8.3 – Preview screen

6. Now, run this in the simulator, and you should see the exact same thing—great job!

7. It's time for a little refactoring now, so make the following highlighted changes to the `ContentView`:

    ```
    var tasks = [Task]()
    var body: some View {
        List(tasks) { task in
            Text(task.description)
        }
    }
    ```

    Here, we're removing our call to our mock data helper as, for production code, we shouldn't be calling this in here. With this change, let's head on over to `PreviewProvider` and make the following highlighted changes there:

    ```
    struct ContentView_Previews: PreviewProvider {
        static var previews: some View {
            ContentView(tasks: MockHelper.getTasks())
        }
    }
    ```

    As our `ContentView` struct now has a non-optional `tasks` variable, it requires us to pass some data in; here, we'll pass in our `MockHelper` function.

8. If not already showing, go ahead and resume the live preview. All being well, everything should be working as expected. However, let's see what happens when we try to run this in the simulator—that's right: no data.

    If you take a closer look at the change we just made, you'll see why we're now only injecting our mock data into our `ContentView` via SwiftUI **PreviewProvider**, so that when our actual app runs, our `tasks` array is empty.

    But this is right, as we'll be pulling our data from a network source in the next recipe. But for now, by injecting the mock data via Preview Provider, we can continue to build our UI long before we build in any networking functionality. So, let's continue.

9. Remember from the previous section how we refactored our `Result` view? We're going to do the same again here for each row in our list view.

Create a new SwiftUI file and call it `ListRowView`, then update the boilerplate code to look like the following:

```
var description: String = ""
var category: String = ""
var body: some View {
    VStack {
        Text(description)
Text(category)
    }
}
```

10. From here, head back over to `ContentView.swift` and make the following changes to the code:

```
var body: some View {
    List(tasks) { task in
        ListRowView(description: task.description, category:
    task.category)
    }
}
```

As we did previously, we're replacing our text view with the view we just created. Click **Resume** to view the live preview and see for yourself.

11. Now, we know that's working, and we want to work on the style of our `ListRowView` a little, so let's head back on over there now and start by updating the Preview Provider so that we can work from there:

```
struct ListRowView_Previews: PreviewProvider {
    static var previews: some View {
        ListRowView(
            description: "Description Field",
            category: "Category Field")
    }
}
```

As you can see from the preceding highlighted code, we've added some mock data for use in our live preview. If this is not already showing, click **Resume** and you should see the following:

Figure 8.4 – Live preview screen with some mock data

12. It works but doesn't look much like a list row, but that's fine—we just need to tell the Preview Provider what we intend to use it for:

```
List {
    ListRowView(description: "Description Field", category:
"Category Field")
}
```

It really is that simple. We just wrap it around a ListView and SwiftUI does the rest, and we can now get to work on decorating our row.

Back in the body of our `ListRowView`, make the following highlighted changes:

```
var body: some View {
    VStack(alignment: .leading) {
        Text(description)
        .font(.title)
        .padding(EdgeInsets(top: 0, leading: 0, bottom: 2,
trailing: 0))
        .foregroundColor(.blue)


        Text(category)
        .font(.title3)
        .foregroundColor(.blue)
    }
}
```

Here, we've added modifiers to our views. Modifiers allow us to decorate and style our views just like we would with properties in UIKit, and each modifier is tied specifically to its type of view.

SwiftUI has gone a little further with some of the modifiers available, giving us a wide variety of options. Let's take the `.font` modifier, for example:

```
.font(.largeTitle) // A font with the large title text style.
.font(.title) // A font with the title text style.
.font(.title2) // Create a font for second level hierarchical
headings.
.font(.title3) // Create a font for third level hierarchical
headings.
.font(.headline) // A font with the headline text style.
.font(.subheadline) // A font with the subheadline text style.
.font(.footnote) // A font with the footnote text style.
.font(.caption) // A font with the caption text style.
.font(.caption2) // Create a font with the alternate caption
text style.
```

The preceding fonts are all available to use straight out of the box; however, if you still want to specify your own font, you can do so by using the following:

```
public static func system(_ style: Font.TextStyle, design: Font.
Design = .default) -> Font
```

13. Let's finish off the base of our app by adding in an image based on the category type:

```
HStack {
    VStack(alignment: .leading) {
        Text(description)
        .font(.title)
        .padding(EdgeInsets(top: 0, leading: 0, bottom: 2,
trailing: 0))
```

```
                .foregroundColor(.blue)

            Text(category)
                .font(.title3)
                .foregroundColor(.blue)
        }

        Spacer()

        Image(systemName: "book")
            .foregroundColor(.blue)
            .padding()
    }
```

Note how, in the preceding code, we've now introduced an HStack and wrapped this around our current VStack, which allows us to add views outside of our original VStack and align them horizontally, just like we've done with the image view.

The use of the Spacer() view in SwiftUI has pushed out our two horizontal views (VStack on the left and Image on the right), so that they act as leading and trailing views to the parent view (the body, in this case).

How it works...

The base of our app is now ready to hook up to an external data source, but first, let's go over how modifiers work and how we can create our very own. Either add the following code to your ListRowView. swift file or create a new file (it's up to you):

```
struct CategoryText: ViewModifier {
    func body(content: Content) -> some View {
        content
            .font(.title3)
            .foregroundColor(.blue)
    }
}
```

Here, we've created a struct called CategoryText that conforms to the ViewModifier protocol. In here, there is a function called body for which we are setting the modifiers of .font and .foregroundColor. These modifiers are available on anything that inherits from View.

Feel free to have a play around with some of the modifiers available. You could add the following and really give the category label a little punch:

```
struct CategoryText: ViewModifier {
    func body(content: Content) -> some View {
```

```
        content
        .font(.footnote)
        .foregroundColor(.blue)
        .padding(4)
        .overlay(
            RoundedRectangle(cornerRadius: 8)
            .stroke(Color.blue, lineWidth: 2)
        )
        .shadow(color: .grey, radius: 2, x: -1, y: -1)
    }
}
```

Let's now add this to our text view:

```
Text(category)
.modifier(CategoryText())
```

We're using `.modifier` in order to call our custom struct; this is good from a readability point of view as it allows you to quickly identify anything that could potentially be custom as opposed to anything that is a system API. However, if you are like me and want it to look just right, simply create an extension of `View`:

```
extension View {
    func styleCategory() -> some View {
        self.modifier(CategoryText())
    }
}
```

Then, use it like this:

```
Text(category)
.styleCategory()
```

You might have noticed we glossed over assigning a specific image to a category. This was with good reason, as this is a perfect opportunity for us to use **SF Symbols** within SwiftUI.

Also available for use with UIKit, SF Symbols works exceptionally well in SwiftUI, especially when used with modifiers like the ones we've just been playing with.

SF Symbols, as the name suggests, is a library of symbols (not images). They are fonts and can be treated just like fonts too, as seen in the following code:

```
Image(systemName: "book")
.font(.system(size: 32, weight: .regular))
.foregroundColor(.blue)
.padding()
```

No need for stretching images or 2x or 3x image assets. SF Symbols will handle this just like having your own vector right in the app, with the added benefit that it's all included in the Swift API without extra assets bulking up your app.

As the name of the symbol is just written in plain text, let's write a little function that works out what we need to display:

```
struct Helper {
    static func getCategoryIcon(category: String) -> String {
        switch category.lowercased() {
        case "shopping":
            return "bag"
        case "health":
            return "heart"
        default:
            return "info.circle"
        }
    }
}
```

In an ideal world, our categories would be enums with `String` values that we could cast to, but for this demo, a basic string match will suffice. Now, replace the static text in the image constructor to call this new static function:

```
Image(systemName: Helper.getCategoryIcon(category: category))
```

The only drawback is the following: is the image you want to use included in the library?

With **iOS 14**, Apple introduced a much wider range of SF Symbols. There's even a Mac app you can download that catalogs all these for you in a nice graphical UI (GUI).

There's more...

We've touched on the Preview Provider a couple of times so far, but this handy little feature of SwiftUI does have a couple more tricks up its sleeve.

By default, the device it will preview on will be that of the one currently selected in Xcode, but if you want to change this, simply add the following:

```
struct ListRowView_Previews: PreviewProvider {
    static var previews: some View {
        List {
            ListRowView(
                description: "Description Field",
                category: "Category Field")
        }
```

```
                .previewDeice(PreviewDevice(rawValue: "iPhone 12 Pro Max"))
                .previewDisplayName("iPhone 12 Pro Max")
        }
    }
```

Let's break the preceding code down:

- `.previewDevice`: This specifies the device you want to use—the raw value string matches that of an internal enum for that specific device (basically the string name as you would see in the simulator list)

- `.previewDisplayName`: This is a custom name given for that device, as shown in the **Live Preview** window

The display name can come in handy for other reasons too, specifically if we have more than one preview running:

```
struct ListRowView_Previews_MockData2: PreviewProvider {
    static var previews: some View {
        List {
            ListRowView(
                    description: "Very Long Description Field, Very
Long Description Field",
                    category: "Very Long Category Field, Very Long
Category Field")
        }
        .previewDevice(PreviewDevice(rawValue: "iPhone 12 Pro"))
        .previewDisplayName("iPhone 12 Pro - Data #2")
    }
}
```

As highlighted, we've created an additional preview to run in our **Live Preview** window, which in turn passes in different data and tests on a different device.

With this, we could create mock data for every condition or style we wanted and have them previewing on all manner of device types, allowing us to test right there without the need to launch each version on a simulator.

SwiftUI is no doubt powerful and brings a more modern approach to building out our UI. However, there are still a number of use cases where we may need to use components and such from UIKit. Luckily for us, Apple has us ready to go with `UIViewRepresentable`, a protocol that we can use to harness UIKit components and return them as SwiftUI views.

A good example would be `UITextView()`, currently not available in SwiftUI or any direct equivalent (although the SwiftUI `TextEditor` component now does a lot of what we want, but is still not a direct replacement as such).

Create a new SwiftUI file and call it `TextView`, then start by pasting the following methods in one by one:

```
struct TextView: UIViewRepresentable {
    @Binding var text: String
    func makeUIView(context: Context) -> UITextView {
        let textView = UITextView()
        return textView
    }
}
```

The `UIViewRepresentable` protocol requires us to conform to certain functions such as `makeUIView()`, which is, in turn, responsible for instantiating the UIKit component we want to wrap.

Next, add the following:

```
struct TextView: UIViewRepresentable {
    // ...
    func updateUIView(_ uiView: UITextView, context: Context) {
        uiView.text = text
    }

    func makeCoordinator() -> Coordinator {
        Coordinator($text)
    }
    // ...
}
```

With `updateUIView()`, we set our `UITextView` instance with whatever we want. Here, we are setting the `text` value from our variable.

Next, we'll add the `makeCoordinator()` function, which returns a `Coordinator` instance, padding in our `@Binding` text field. The best way to think about `Coordinator` is as a way of handling the delegate methods we might use for our UIKit component. Add in the following, and this should make more sense:

```
struct TextView: UIViewRepresentable {
    // ...
    class Coordinator: NSObject, UITextViewDelegate {
        var text: Binding<String>
        init(_ text: Binding<String>) {
            self.text = text
        }

        func textViewDidChange(_ textView: UITextView) {
```

```
                    self.text.wrappedValue = textView.text
            }
        }
    }
```

See how our `Coordinator` instance conforms to `UITextViewDelegate`, and we have `textViewDidChange()` in there. As our `text` variable being passed in is a `Binding` string, changes made will reflect in the delegate method being called, just as they would in UIKit:

```
@State var textViewString = ""
TextView(text: $textViewString)
```

In order to call this, we would simply add this as we would any other SwiftUI view.

See also

- SF Symbols: `https://developer.apple.com/design/human-interface-guidelines/sf-symbols/overview/`

- `UIViewRepresentable`: `https://developer.apple.com/documentation/swiftui/uiviewrepresentable`

Combine and data flow in SwiftUI

For many years, the reactive programming stream has played a big part in development architecture in terms of iOS and macOS. You may have heard of **RxSwift** and **RxCocoa**, a massive community committed to the reactive stream that allows for asynchronous events to be processed.

If you are not familiar with the terminology of **Rx** or **reactive programming**, you may have seen the use of `Publishers`, `Subscribers`, and `Operators` in your code base. If you have, then you've most likely been subject to reactive programming at some point.

In this section, we are going to take a look at Apple's offering for reactive programming, called **Combine**. Introduced alongside SwiftUI at WWDC 2019, Combine is the perfect accompaniment for the new layout and structure of SwiftUI (although not bound solely to SwiftUI). We'll take a look at how we can create a seamless flow of data from an online resource, right up to our UI layer.

Getting ready

For this section, you'll need the latest version of Xcode from the Mac App Store and the project from the previous section.

How to do it...

1. First, we'll start by updating our `Task` model to a class, by making the following highlighted changes:

```
class Task: Identifiable {
    var id = UUID()
    let response: TaskResponse

    init(taskResponse: TaskResponse) {
        self.response = taskResponse
    }

    var category: String {
        return response.category ?? ""
    }

    var description: String {
        return response.description ?? ""
    }
}
```

2. Here, we've converted our model to a class (more on that later), and have added a custom initializer and a couple of computed properties too.

    We've also added a variable of type `TaskResponse`, so let's go ahead and create that now in a new file:

```
struct TaskResponse: Codable {
    let category: String?
    let description: String?
}
```

3. Here, we have a basic codable response. Now, for a bit of boilerplate networking code, create a new file called `NetworkManager.swift`, and add the following code into it:

```
class NetworkManager {
    static func loadData(url: URL, completion: @escaping
([TaskResponse]?) -> ()) {
        URLSession.shared.dataTask(with: url) { data,
response, error in
            guard let data = data, error == nil else {
                completion(nil)
                return
            }
            if let response = try? JSONDecoder().
decode([TaskResponse].self, from: data) {
```

```
                        DispatchQueue.main.async {
                            completion(response)
                        }
                    }
                }.resume()
            }
        }
```

4. Here, we have a basic implementation of `URLSession`, which is being passed as a **Uniform Resource Locator (URL)**, parsing the **JavaScript Object Notation (JSON)** response into a codable object (our `TaskResponse` model). The function we've created has a completion handler that returns an array of our `TaskResponse` model, should the response and decoding be successful.

Next, create a file called `TaskViewModel`, and add in the following code:

```
class TaskViewModel: ObservableObject {
    init() {
        getTasks()
    }

    @Published var tasks = [Task]()

    private func getTasks() {
        guard let url = URL(string: "https://raw.
githubusercontent.com/PacktPublishing/Swift-Cookbook-Third-
Edition/main/Chapter%208/TaskResponse.json") else {
            return
        }
        NetworkManager.loadData(url: url) {
taskResponse in
            if let taskResponse = taskResponse {
                self.tasks = taskResponse.
map(Task.init)
            }
        }
    }
}
```

In the preceding code, I've highlighted some areas of interest. First is how we've conformed our class to `ObservableObject`—this is required as our `tasks` variable has the `@Published` wrapper and will be looking for changes as and when they occur.

Next is the local URL we're passing into `NetworkingManager`—this is the address for a JSON file hosted in this chapter's repo.

5. Now, head on back over to our `ContentView.swift` file and make the following highlighted changes:

```
struct ContentView: View {
    @ObservedObject var model = TaskViewModel()
    var body: some View {
        List(model.tasks) { task in
            ListRowView(
                    description: task.description,
                    category: task.category)
        }
    }
}
```

We've now renamed our `tasks` variable to `model`, and this in turn has now created a new `TaskViewModel()` instance.

6. As we've updated a few things here, the structure of how we inject our mock data will need adjusting too, so make the following highlighted changes to our `MockHelper` function:

```
var task = [Task]()
task.append(Task(taskResponse: TaskResponse(category: "Get
Eggs", description: "Shopping")))
task.append(Task(taskResponse: TaskResponse(category: "Get
Milk", description: "Shopping")))
task.append(Task(taskResponse: TaskResponse(category: "Go for a
run", description: "Health")))

let taskViewModel = TaskViewModel()
taskViewModel.tasks = task
return taskViewModel
```

It's time to see the magic happen. Launch the app, and if all's going well, you should see something like this:

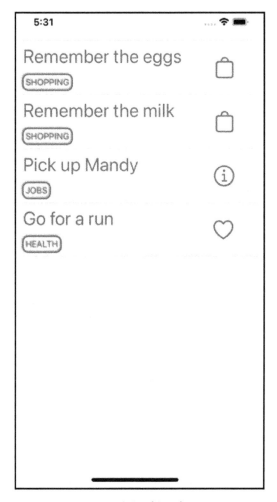

Figure 8.5 – Launching the app

Here is a simple yet exceptionally effective demonstration of how Combine can and should be used within SwiftUI. Let's take a look now at how all this actually works.

How it works...

Let's start at ContentView and work our way back:

```
@ObservedObject var model = TaskViewModel()
```

Two things to note here—our model is that of @ObservedObject, meaning any changes made to this model will result in an update being fired and thus forcing a refresh of our UI (just like we saw with @State earlier).

Next, we're instantiating `TaskViewModel()` when `ContentView` is rendered. Let's dive into `TaskViewModel` and see why:

```
class TaskViewModel: ObservableObject {
    init() {
        getTasks()
    }
    @Published var tasks = [Task]()
    // ...
}
```

We already touched earlier on our class conforming to `ObservableObject`. This is what allows us to use `@ObservedObject` when declaring this back over in `ContentView` (we're creating a data flow connection, so to speak).

Notice here that we've also added a call to our `getTasks()` function so that when we initialize the class (back over in `ContentView`), we'll kick off a networking request to get a list of tasks.

If we now have a quick look inside our `getTasks()` function, you'll see that once we get a successful response, we assign this to our `@Published tasks` variable:

```
NetworkManager.loadData(url: url) { tasksResponse in
    if let tasksResponse = tasksResponse {
        self.tasks = tasksResponse.map(Task.init)
    }
}
```

As soon as the variable is updated, our `Observable` object class lets anything listening know about a change (`@ObservedObject` in `ContentView`, for example).

If you think back to how `UITableView` works, if there are any updates or changes to the data source, we then have to call `UITableView.reloadData()` manually, and within our UI layer.

With this approach, everything has been handled the way it should be and is in the right place, passing data changes from the source of truth up to the UI layer.

This is just a basic example of how Combine works, specifically in tandem with SwiftUI. In *Chapter 9, Getting to Grips with Combine*, we will go deeper into how Combine works and even compare it to the Delegate pattern!

See also

- Swift Combine: `https://developer.apple.com/documentation/combine`

Getting to Grips with Combine

9

Reactive programming is a paradigm that involves designing systems that respond to changes in the input data and automatically update their output. In reactive programming, changes to the data are modeled as a stream of events, and operations are performed on these streams to produce new streams of updated data. This allows developers to write more concise and maintainable code by abstracting complex logic into simple operations on streams.

Combine is a reactive programming framework developed by Apple for use in iOS, iPadOS, and macOS development. It provides a set of APIs for processing values over time as well as handling events and data changes in a declarative and functional manner. Combine makes it easy to write reactive code by providing a wide range of operators and publishers that can be used to manipulate streams of data. It also provides a unified mechanism for handling errors and cancellations, making it easier to write robust and reliable reactive code.

By the end of this chapter, you will have learned how to leverage Combine to build reactive apps that read more declaratively and function more asynchronously.

In this chapter, we will cover the following recipes:

- Using Reactive Streams
- Understanding Observable Objects
- Understanding publishers and subscribers
- Combine versus Delegate pattern

Technical requirements

You can find the code files present in this chapter on GitHub at `https://github.com/PacktPublishing/Swift-Cookbook-Third-Edition/tree/main/Chapter%209`.

Using Reactive Streams

In this first recipe, we're going to create a simple SwiftUI view to demonstrate a simple reactive flow. Considering SwiftUI and Combine were released at the same time, it only makes sense to review reactive programming using Apple's first-party frameworks built specifically for it!

Getting ready

For all the recipes in this chapter, you'll need the latest version of Xcode available from the Mac App Store.

How to do it...

With Xcode open, let's get started:

1. Create a new project in Xcode. Go to **File** | **New** | **Project** | **iOS App**. Be sure to choose SwiftUI for the interface.

2. First, create a view called FetchView that simply holds a Binding property and a body that displays that property's value:

   ```
   struct FetchView: View {
       @Binding var nameDownStream: String

       var body: some View {
           VStack(alignment: .center) {
               Text("Go fetch the 🎾\(nameDownStream)!")
           }
       }
   }
   ```

3. In ContentView, first add a State property:

   ```
   @State private var nameStream = ""
   ```

4. Replace the contents of body with Text, TextField, Divider, and our new FetchView, all within VStack:

   ```
   VStack {
       Text("Your Pets Name is: \(nameStream)")
       TextField("Enter name", text: $nameStream)
           .textFieldStyle(RoundedBorderTextFieldStyle())
       Divider()
       FetchView(nameDownStream: $nameStream)
   }
   ```

As you'll see in *Figure 9.1*, we have a very basic app layout that uses our `nameStream State` in multiple sub-views.

Figure 9.1 – A simple yet reactive app!

How it works...

This is a very simple recipe, but it proves a point: reactive programming, especially using Combine and SwiftUI, makes things simpler!

In `ContentView`, the `TextField` view is bound to `@State` property text, which will automatically update any view that points to that property when the user enters text into the field. The `Text` view is then used to display the entered text, which is updated in real time as the user types. This is an example of a reactive stream where the changes in the input are automatically reflected in the output, without the need for manual intervention.

To further demonstrate a reactive flow, we also added `FetchView`. By adding a property wrapped with `Binding`, we essentially declare that we want to **subscribe** (a key term in reactive programming) to the updates of whatever `State` property is assigned:

```
@Binding var nameDownStream: String
```

Once we set the `Text` view to use that binding, that view will now reflect the updated value in real time, even though it's in a different view.

The key takeaway is that unlike in most other paradigms, there was no need for any functions or added logic to take the input and pass it along to the appropriate UI destinations. We simply set the property in the places we expected to input into and output from it. When we run our app, we can begin typing, and our UI reacts accordingly.

The reactive programming paradigm and the Combine framework make it easy to build reactive applications in SwiftUI, allowing developers to write more concise and maintainable code. The combination of SwiftUI and Combine, especially its more advanced features that we'll learn about in the upcoming recipes, provides a powerful and streamlined way to build reactive UIs in our apps.

See also

For more information on `State` and `Binding`, refer to `https://developer.apple.com/documentation/swiftui/managing-user-interface-state`.

Understanding Observable Objects

Now that we understand reactive flows, we can dive a little deeper into more complex situations. In this recipe, we will look at **Observable Objects** and see how they can help make our apps more reactive to data structures.

How to do it...

Let's start where we left off in our previous recipe:

1. First, create a new `class` that conforms to `ObservableObject` and will hold all of our pet's information:

```
class ObservablePet : ObservableObject {
    @Published var name = «»
    @Published var age = «»
    @Published var breed = «»

    init() { }
}
```

2. Next, in `ContentView`, change our `State` property to a `StateObject` property:

```
@StateObject var pet = ObservablePet()
```

3. In body, we want to account for our pet's attributes, so add more `Text` and `TextField` views, pointing to their respective properties:

```
Text("Your Pets Name is: \(pet.name)")
Text("Your Pets Age is: \(pet.name)")
Text("Your Pets Breed is: \(pet.name)")

TextField("Enter name", text: $pet.name)
    .textFieldStyle( .roundedBorder)
TextField("Enter age", text: $pet.age)
    .textFieldStyle( .roundedBorder)
TextField("Enter breed", text: $pet.breed)
    .textFieldStyle( .roundedBorder)
```

4. Over in `FetchView`, swap out our `Binding` for an `ObservedObject` property:

```
@ObservedObject var petDownStream: ObservablePet
```

5. And let's update our `Text` to show off all our data:

```
Text("\(petDownStream.name), a \(petDownStream.age) year old \
(petDownStream.breed) will fetch the 🎾")
```

6. Lastly, update our declaration of `FetchView` back in `ContentView`:

```
FetchView(petDownStream: pet)
```

As visible in *Figure 9.2*, the app now shows more information about the pet.

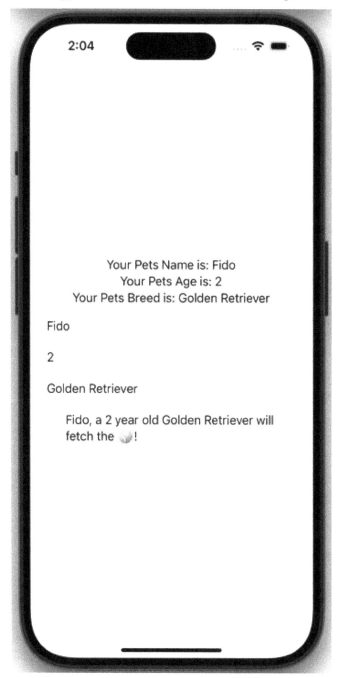

Figure 9.2 – Now we know more about Fido!

How it works...

This recipe works very similarly to the last one, but with a few key differences that begin to open up possibilities for us. Let's examine our new class `ObservablePet`:

```
class ObservablePet : ObservableObject {
    @Published var name = «»
    @Published var age = «»
    @Published var breed = «»

    init() { }
}
```

The first thing to notice is that it looks like any other basic class. However, by conforming to `ObservableObject`, we now attach it to Combine and signal that this is a reactive class. We can also, then, specify which properties in our class we expect to be reactive by adding the `Published` property wrapper. Not every property of a class may need to be published, so using the wrapper allows us to be specific.

Back in our views, two major differences are that `State` was swapped out for `StateObject`, and `Binding` for `ObservableObject`. These are the appropriate wrappers when working with properties that are not basic types, such as `ObservableObject`, and have published properties within them.

From there, we simply expanded our UI in both Views to account for our new data points, publishing to and subscribing to each point where needed.

See also

For more information on observables, refer to `https://developer.apple.com/documentation/combine/observableobject`.

Understanding publishers and subscribers

You may recall seeing the terms `publish` and `subscribe` in the previous recipes. In Combine, these are the core concepts working behind the scenes to create reactive data streams.

When you think about it, it reveals a lot about how it all works. A publisher *broadcasts* information, and subscribers choose to *tune in and listen* for relevant data to react to.

We'll explore this idea in this recipe where we'll publish the results of a network call for `Puppy` information, and then subscribe to that publisher with some instructions on what to do.

How to do it...

We will start with a new playground:

1. Create a new playground in Xcode. Go to **File** | **New** | **Playground** | **iOS**.

2. For this, we will need to import both Foundation and Combine at the top.

3. First, let's make our Puppy struct. We expect to read three properties from JSON on our network call:

   ```
   struct Puppy: Decodable {
           let id: Int
           let name: String
           let breed: String
   }
   ```

4. Then, let's set the URL where we'll find our data. A JSON file with Puppy information is already available in the repo for this recipe, so you can just use the following:

   ```
   let url = URL(string: "https://raw.githubusercontent.com/
   PacktPublishing/Swift-Cookbook-Third-Edition/main/Chapter%209/
   Chapter%209%20-%203/Pups.json")!
   ```

5. Next, it's time to create our publisher. Fortunately, one is already provided as part of URLSession, so we can use that as our base:

   ```
   let publisher = URLSession.shared
           .dataTaskPublisher(for: url)
   ```

6. Then we'll add a few operators right below dataTaskPublisher to process our data before publishing:

   ```
           .map { $0.data }
           .decode(type: [Puppy].self, decoder: JSONDecoder())
           .eraseToAnyPublisher()
   ```

7. Now for our subscriber! To do this, we'll call sink on publisher:

   ```
   let subscriber = publisher
           .sink(receiveCompletion: { completion in
                   // TODO
           }, receiveValue: { pups in
                   // TODO
           })
   ```

8. Notice the two closures. For the first, `receiveCompletion`, we'll add a simple switch to handle failure and finished states:

```
let subscriber = publisher
    .sink(receiveCompletion: { completion in
        switch completion {
        case .failure(let error):
            print("Error: \(error.
localizedDescription)")
        case .finished:
            print(«Completed»)
        }
    }, receiveValue: { pups in
        // TODO
    })
```

9. For the second closure, we hope to receive our Puppy information! We'll simply print out that information, giving each Puppy a treat as we share their info:

```
let subscriber = publisher
    .sink(receiveCompletion: { completion in
        switch completion {
        case .failure(let error):
            print("Error: \(error.localizedDescription)")
        case .finished:
            print(«Completed»)
        }
    }, receiveValue: { pups in
        for pup in pups {
            print("Feeding \(pup.name) the \(pup.breed) a
treat!")
        }
    })
```

10. Lastly, we run our playground and should expect to feed three happy pups, as shown in *Figure 9.3*!

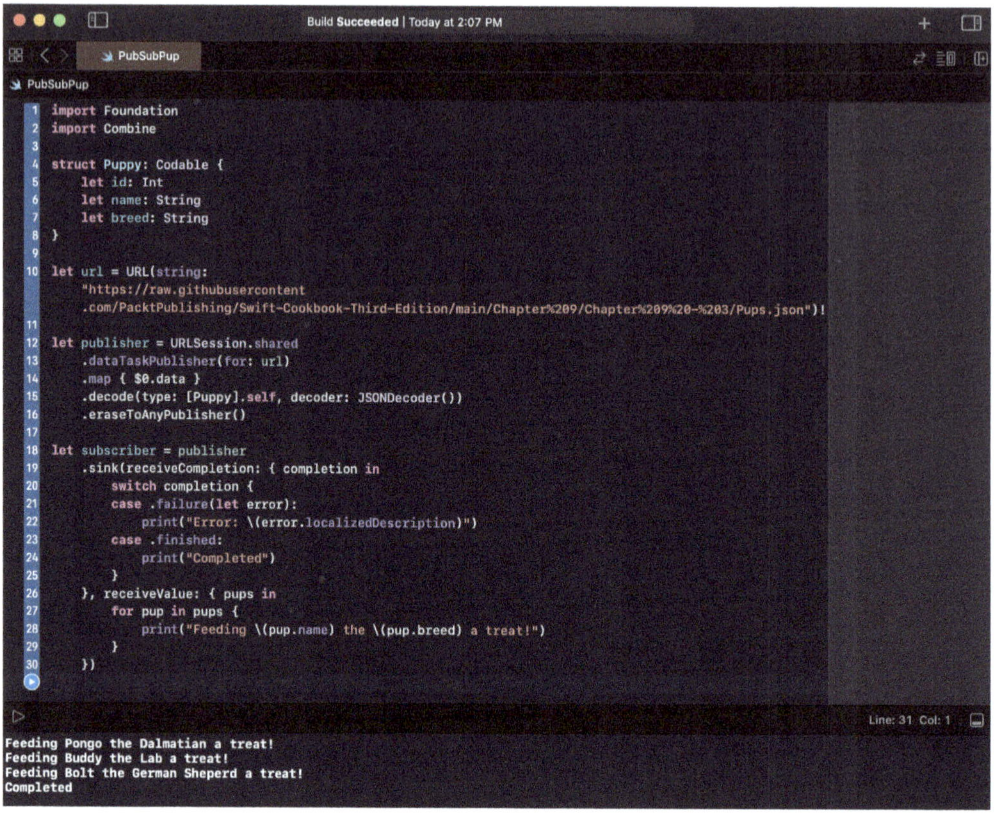

Figure 9.3 – As our publisher receives values, we feed our pups!

How it works...

Setting up our `Puppy` struct and URL should be recognizable by now, so let's jump right into how the publisher works.

With Combine, Apple has automatically provided access to publishers straight from commonly used objects found in Foundation. Fortunately for us, this includes a publisher based on `URLSession` called `dataTaskPublisher`:

```
let publisher = URLSession.shared
    .dataTaskPublisher(for: url)
```

If we were to dig into `dataTaskPublisher` a bit, we'd find it conforms to the `Publisher` protocol, which requires a defined output (in this case, a tuple of `Data` and `URLResponse`), failure (`URLError`), and a `receive` function (where our subscriber will become attached). These make up a core `Publisher`.

Now, when `dataTaskPublisher` is run and our data is received, we could publish the data right there and then, but it would be in a raw state. To format it into a `Puppy` struct, we call a series of publisher operators. They help process information from upstream down until we're satisfied with the state we expect it to be in.

In our case, since we're receiving JSON and `Puppy` is codable, we want to first map the data from the response, then decode the data, and, lastly, create our `Publisher` as `AnyPublisher`, erased of any of its types:

```
let publisher = URLSession.shared
        .dataTaskPublisher(for: url)
        .map { $0.data }
        .decode(type: [Puppy].self, decoder: JSONDecoder())
        .eraseToAnyPublisher()
```

Notice that we can call operators one after another. This allows our `Publisher` and its operators to be sequential and readable.

From our publisher, we now want to create a subscriber or an object that will listen to what `Publisher` sends out and carry out certain instructions when it does receive data. To do this, we can simply call `sink` on our `publisher`:

```
let subscriber = publisher
    .sink(receiveCompletion: { completion in
            // TODO
    }, receiveValue: { pups in
            // TODO
    })
```

The `sink` method simply creates and then attaches a new `Subscriber` to a `Publisher`. The `Subscriber` protocol, in correlation to a `Publisher`, requires `input` and `failure`. Because `sink` is actually a publisher operator itself, it can determine the types for `input` and `failure` automatically, creating a perfect `Subscriber`.

The only things left to define are the two completions offered by `sink`. The first, `receiveCompletion`, simply provides the opportunity to handle different completion states:

```
receiveCompletion: { completion in
        switch completion {
        case .failure(let error):
            print("Error: \(error.localizedDescription)")
        case .finished:
            print(«Completed»)
        }
    },
```

The second, `receiveValue`, is of more interest to us, as it allows us to take the data we're expecting (an array of `Puppy`) and do whatever we plan on using the data for (such as feeding our puppies treats!):

```
receiveValue: { pups in
            for pup in pups {
                    print("Feeding \(pup.name) the \(pup.breed) a
treat!")
            }
        })
```

Upon execution of this recipe, it should run rather quickly. However, behind the scenes, there are a few things to note.

First, our `publisher` only cares about sending information to its subscribers. Even then, it's not necessarily concerned about what our subscribers do, but rather that when it has something to send, it knows who to send it to.

Second, our `subscriber` is patiently waiting for information before doing anything. Its purpose is to wait for information, and when (and only when) it receives any, then it will react. This is great because we don't have to anticipate or hold up anything should we, for instance, have a slow or bad network and our network call doesn't produce data so quickly.

See also

Use these links to learn more:

- Publishers: `https://developer.apple.com/documentation/combine/publisher`

- Subscribers: `https://developer.apple.com/documentation/combine/subscriber`

- Publisher operators: `https://developer.apple.com/documentation/combine/publishers-decode-publisher-operators`

Combine versus Delegate pattern

When Swift was first released, it strongly urged **protocol-oriented programming**, which promoted the **Delegate** pattern. The way the delegate pattern works is one object would act on behalf of, or in coordination with, another object. In iOS development, delegate patterns have been used to handle events and user interactions for years.

That is, until reactive programming came into view, especially with Combine. We'll do a simple comparison in playgrounds using `Timer` so we can observe the evolution between the two.

How to do it...

We will start with a new playground:

1. Create a new playground in Xcode. Go to **File** | **New** | **Playground** | **iOS**.

2. On our first page, we'll start with the Delegate approach. First, we'll create our delegate protocol:

```
protocol TimerDelegate: AnyObject {
        func timerEventReceived()
}
```

3. Create a new class that will feed our delegate timer events:

```
class TimerSender {
        var timer = Timer()
        weak var delegate: TimerDelegate?
}
```

4. Add a function to be called at every event, and trigger the delegate call within it:

```
@objc func timerEvent() {
        delegate?.timerEventReceived()
}
```

5. Lastly, we'll set up init to automatically start sending timer events:

```
init() {
        self.timer = Timer.scheduledTimer(timeInterval: 1, target:
self, selector: #selector(timerEvent), userInfo: nil, repeats:
true)
}
```

6. Now we'll create another class that conforms to our delegate protocol and has a simple print statement in its conformance:

```
class TimerReceiver: TimerDelegate {
        func timerEventReceived() {
                print("Timer event received")
        }
}
```

7. Lastly, we create an instance of both of our classes and then set our delegate. When we run the playground, you'll start to see the expected print statements in the console log!

```
var receiver = TimerReceiver()
var sender = TimerSender()
sender.delegate = receiver
```

8. Let's now do this using Combine! Create a new playground page by going to **File** | **New** | **Playground Page**.

9. First, be sure we're importing Combine and Foundation.

10. Next, set up a timer, this time capturing its publisher, which is already provided to us in Foundation:

```
let timer = Timer.publish(every: 1, on: .main, in: .default)
```

11. We then subscribe to the publisher, where we'll simply print on every event:

```
let subscription = timer
        .map { _ in
            print("Timer event received")
        }
        .sink { _ in   }
```

12. Lastly, we start the timer as follows:

```
timer.connect()
```

How it works...

We'll break this up into two parts. First, in our Delegate example, we set up our delegate protocol, `TimerDelegate`, with a single function stub (`timerEventReceived`) to pass our events through. This allows `TimerSender` to be able to trigger delegates:

```
delegate?.timerEventReceived()
```

Our delegate, `TimerReceiver`, conforms to the `TimerDelegate` protocol and simply implements our stub with a print statement.

In our Combine example, things are much simpler, especially since Foundation provides a publisher on `Timer`. From there, it's incredibly easy to subscribe with a basic print statement that's triggered on every event.

While these are simple examples, let's dive into the differences. Our Delegate example required a protocol, two classes, and instances of those classes for us to achieve a data stream on our `Timer`. On the other hand, through Combine, all we needed were two basic instances of `Publisher` (which happened to already be available to us from `Timer` directly) and a `Subscriber` instance, which we got from `Publisher`. That is significantly less code to compile, read, and instantiate to get the same result.

What better way to end our journey with Combine than to show how it has evolved Swift code? Combine bringing first-party support for reactive programming to Swift is a leap forward for the language and our apps!

10
Using CoreML and Vision in Swift

The Swift programming language has come a long way since its first introduction, and in comparison to many other programming languages, it's still well within its infancy.

However, with this in mind, with every release of Swift and its place in the open source community, we've seen it grow from strength to strength over such a short period of time. One of these core strengths is machine learning.

In this chapter, we're going to look at Apple's offering for machine learning – CoreML – and how we can build an app using Swift to read and process machine learning models, giving us intelligent image recognition.

We'll also take a look at Apple's Vision framework and how it works alongside CoreML to allow us to process video being streamed to our devices in real time, recognizing objects on the fly.

This will lay the foundation for bringing machine learning into your apps and their features, a step into the future of personalized and enhanced user experiences.

In this chapter, we will cover the following recipes:

- Building an image capture app
- Using CoreML models to detect objects in images
- Building a video capture app
- Using CoreML and the Vision framework to detect objects in real time

Technical requirements

You can find the code files present in this chapter on GitHub at `https://github.com/ PacktPublishing/Swift-Cookbook-Third-Edition/tree/main/Chapter%2010/ Chapter%2010%20-%20Core%20ML`.

Building an image capture app

In this first recipe, we're going to create an app that captures either an image from your camera roll or an image taken from your camera. This will set up our iOS app ready for us to incorporate CoreML to detect objects in our photos.

Getting ready

For this recipe, you'll need the latest version of Xcode available from the Mac App Store.

How to do it...

With Xcode open, let's get started:

1. Create a new project in Xcode. Go to **File** | **New** | **Project** | **iOS App**.

2. In `Main.storyboard`, add the following:

    I.   Add `UISegmentedControl` with two options (`Photo / Camera Roll` and `Live Camera`).

    II. Next, add a `UILabel` view just underneath.

    III. Add a `UIImageView` view beneath that.

    IV. Finally, add a `UIButton` component.

3. Space these accordingly using `AutoLayout` constraints with `UIImageView` being the prominent object:

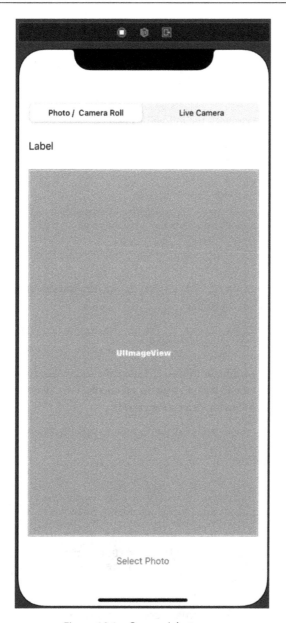

Figure 10.1 – Camera/photo app

4. Once we have this in place, let's hook these up to our `ViewController.swift` file:

```
@IBOutlet weak var imageView: UIImageView!
@IBOutlet weak var labelView: UILabel!
@IBAction func onSelectPhoto(_ sender: Any)
```

> **Note**
>
> Take note that in the preceding code block, we have two outlets, using the `IBOutlet` keyword, and one action, using the `IBAction` keyword. We don't need an outlet for the `UIButton`, as we only care about its action at this stage.

5. Next, populate `IBAction` with the following code:

```swift
@IBAction func onSelectPhoto(_ sender: Any) {
    let picker = UIImagePickerController()
    picker.delegate = self
    picker.allowsEditing = false
    picker.sourceType = UIImagePickerController.
isSourceTypeAvailable(.camera) ? .camera : .photoLibrary
    present(picker, animated: true)
}
```

6. Now, let's create an extension of `UIViewController`. You can do this at the bottom of the `ViewController` class if you like:

```swift
extension ViewController: UIImagePickerControllerDelegate,
UINavigationControllerDelegate
```

7. Our extension needs to conform to the `UIImagePickerControllerDelegate` and `UINavigationControllerDelegate` protocols. We can now go ahead and populate our extension with the following delegate method:

```swift
func imagePickerControllerDidCancel(_ picker:
UIImagePickerController) {
    dismiss(animated: true, completion: nil)
}

func imagePickerController(_ picker: UIImagePickerController,
didFinishPickingMediaWithInfo info: [UIImagePickerController.
InfoKey : Any]) {
    guard let image = info[UIImagePickerController.InfoKey.
originalImage] as? UIImage else { return }
    imageView.image = image
    labelView.text = "This is my image!"
    dismiss(animated: true, completion: nil)
}
```

8. Before we go any further, we'll need to add a couple of lines to `info.plist`:

```
NSCameraUsageDescription
NSPhotoLibraryUsageDescription
```

9. Add these in with the following string description: Chapter 10 wants to detect cool Stuff. This is an iOS security feature that will prompt the user when any app/code tries to access the camera, photo library, or location services. Failure to add this could result in an app crash.

> **Note**
>
> For our app, we can add whatever we want, but for a production app, make sure the text you enter is useful and informative to the user. Apple will check this when reviewing your app and has been known to potentially block a release until this is resolved.

Go ahead and run your code, and then launch the app. Once the app has been launched, you will notice a prompt that asks for permission to access the camera. This prompt should only appear once, as the permission can be changed in the device settings at a later point if required. After acknowledging the prompt, one of the following things should happen:

- If you are running the app from the simulator, our UIButton press should present the photo picker (along with the default images supplied by the iOS simulator)

- If you are running from a device, then you should be presented with the camera view, allowing you to capture a photo

Either way, whether a photo was selected or a picture was taken, the resulting image should show in UIImageView!

How it works...

Let's step through what we've just done. We'll begin at IBAction and have a look at the UIPickerView view we've created:

```
let picker = UIImagePickerController() // 1
picker.delegate = self // 2
picker.allowsEditing = false// 3
picker.sourceType = UIImagePickerController.isSourceTypeAvailable(.
camera) ? .camera : .photoLibrary // 4
present(picker, animated: true) // 5
```

Let's go through this one line at a time:

1. We instantiate an instance of UIImagePickerController – an available API that will allow us to choose an image based on a specific source.

2. We set the delegate as self, so we can harness any results or actions caused by UIImage-PickerController.

3. We set allowEditing to false, which is used to hide controls when the camera is our source.

4.   In this instance, we set the source type based on whether the camera is available or not (so it works well with the simulator).

5.   Finally, we present our view controller.

Now, let's take a look at our delegate methods:

```
func imagePickerControllerDidCancel(_ picker: UIImagePickerController)
func imagePickerController(_ picker: UIImagePickerController,
didFinishPickingMediaWithInfo info: [UIImagePickerController.InfoKey:
Any])
```

The first method is pretty self-explanatory; `imagePickerControllerDidCancel` handles any instances where `UIImagePickerController` is canceled by the users. In our case, we just dismiss the instance returned – job done!

`didFinishPickingMediaWithInfo` is where interesting things happen. Notice how we are given a dictionary of information in our response. Here, we have various segments of information. The one we are looking for is under the `UIImagePickerController.InfoKey.originalImage` key. This gives us an image of what we've just selected, in the form of `UIImage`, allowing us to assign this straight back to `UIImageView`.

Now that we've got an app that allows us to take or choose a photo, we can apply it to some real work with the power of CoreML and object detection.

There's more...

A quick note to mention: you'll also have noticed that we were required to conform our extension to `UINavigationControllerDelegate`. This is required by iOS to allow `UIImageController` to be handled and presented correctly from its *presenting* stack (`ViewController`, in our instance).

See also

For more information on `UIImagePickerController`, refer to `https://developer.apple.com/documentation/uikit/uiimagepickercontroller`.

Using CoreML models to detect objects in images

In this recipe, we'll take the app we just built and incorporate the CoreML framework in order to detect objects in our images.

We'll also take a look at the generated CoreML models available for us to use and download directly from Apple's Developer portal.

Getting ready

For this recipe, you'll need the latest version of Xcode available from the Mac App Store.

Next, head on over to the Apple Developer portal at the following address: `https://developer.apple.com/machine-learning/models/`.

Here, you will find out a little bit more about the models available for us to download and use in our Xcode project.

You'll notice there are options for image models and text models. For this recipe, we're going to be using image models, specifically one called **Resnet50**, which uses a residual neural network that attempts to identify and classify what it perceives to be the dominant object in an image.

> **Note**
>
> For more information on the different types of machine learning models, see the links in the *See also* section at the end of this recipe.

From the URL provided in this section, download the **Resnet50.mlmodel** (32-bit) model.

Once downloaded, add this to your Xcode project by simply dragging it into the File Explorer tree in our previous app.

How to do it...

Let's make a start where we left off in our previous project:

1. With everything in place, head back into `ViewController.swift` and add the following global variable and addition to our `viewDidLoad()` function:

    ```swift
    var model: Resnet50!
    override func viewDidLoad() {
        super.viewDidLoad()
        model = try? Resnet50(configuration: .init())
    }
    ```

2. Now, head on over to the sample project and obtain a file called `ImageHelper.swift`; add this to our project. Once this has been added, we'll head on back over to our `didFinishPickingMediaWithInfo` delegate and expand on this a little further.

3. Add in the following changes:

    ```swift
    guard let image = info[UIImagePickerController.InfoKey.
    originalImage] as? UIImage else {
        return
    }
    ```

```
let (newImage, pixelBuffer) = ImageHelper.
processImageData(capturedImage: image)
imageView.image = newImage
var imagePredictionText = "no idea... lol"
guard let prediction = try? model.prediction(image:
pixelBuffer!) else {
    labelView.text = imagePredictionText
    dismiss(animated: true, completion: nil)
     return
}
imagePredictionText = prediction.classLabel
labelView.text = "I think this is a \(imagePredictionText)"
dismiss(animated: true, completion: nil)
```

With everything in place, run the app and select a photo. As long as you didn't point it at a blank wall, you should begin to receive predictions.

With all that in place, let's break down the changes we just made to understand what just happened a little more.

How it works...

The first thing is to take a look at the following line we added:

```
ImageHelper.processImageData(capturedImage: image)
```

Here, we added a call to a helper method we took from our sample project. This helper contains the following two functions:

```
static func processImageData(capturedImage: UIImage) -> (UIImage?,
CVPixelBuffer?)
static func exifOrientationFromDeviceOrientation() ->
CGImagePropertyOrientation
```

These functions and what they do are a little out of the scope of this book, and this chapter in particular. However, at a very high level, the first function, `processImageData()`, takes an instance of `UIImage` and transforms this to `CVPixelBuffer` format.

This essentially returns the `UIImage` object back to the raw format that it was captured in (`UIImage` is merely a `UIKit` wrapper for our true raw image).

During this process, we need to flip the orientation too, as with all captured images. This is almost certainly in landscape mode (and more often than not, you've taken a picture or selected a photo in portrait mode).

Another reason for performing this is that our Resnet50 model is trained to observe images at only 224 x 224. So, we need to readjust the captured image to this size.

> **Note**
>
> If you need more information on the model you have in your project, simply select the file in the File Explorer and view the details in the main window. From here, the **Predictions** tab will give you all the details you need about the input file required.

So, with our helper function implemented, we receive a new `UIImage` object (modified to our new spec) and the image in the `CVPixelBuffer` format, all ready to pass over to CoreML for processing.

Now, let's take a look at the following code:

```
guard let prediction = try? model.prediction(image: pixelBuffer!) else
{
    labelView.text = imagePredictionText
    dismiss(animated: true, completion: nil)
    return
}
imagePredictionText = prediction.classLabel
```

First is our `prediction()` function call on our model object. Here, we pass in our image in the `CVPixelBuffer` format we got back from our helper method earlier. From this, wrapped in a `try` statement, CoreML will now attempt to detect an object in the photo. If successful, we'll exit our `guard` statement gracefully and be able to access the properties available in our `prediction` variable.

If you take a look at the properties available in our Resnet50 model, by selecting the file and opening the **Predictions** tab, you'll see the various output options we have:

```
.classLabel
.classLabelProbs
```

The class label we've already seen, but the class label probability will return a dictionary of the most likely category for our image with a value based on a confidence score (an automatically generated percentage that the model has predicted correctly).

Each model will have its own set of properties based on its desired intention and how it's been built.

There's more...

At the beginning of this recipe, we obtained a model that allowed us to detect objects in our images. Touching on this subject a little more, a model is a set of data that has been trained to identify a pattern or characteristics of a certain description.

For example, we want a model that detects cats, so we train our model by feeding it images of around 10,000 various pictures of cats. Our model training will identify features and shapes common to each other and categorize them accordingly.

When we then feed our model an image of a cat, we hope that it is able to pick up those categorized features within our image and successfully identify the cat.

The more images you train with, the greater the performance; however, that depends on the integrity of the images, too. Training with the same image of a cat (just in a different pose) 1,000 times might give you the same results as if you take 10,000 images of the same cat (again in a different pose).

The same goes the other way, too; if you train with 500,000 images of a panther and then 500,000 images of a kitten, it's just not going to work.

See also

For more information, please refer to the following link within the Apple CoreML documentation: `https://developer.apple.com/documentation/coreml`.

Building a video capture app

So, what we have seen so far of CoreML is pretty neat, to say the least. But taking a look back over this chapter so far, we have probably spent more time building our app to harness the power of CoreML than actually implementing it.

In this section, we're going to take our app a little further by streaming a live camera feed that, in turn, will allow us to intercept each frame and detect objects in real time.

Getting ready

For this section, you'll need the latest version of Xcode available from the Mac App Store.

Please note that, for this section, you'll need to be connected to a real device for this to work. Currently, the iOS simulator does not have a way to emulate the front or back camera.

How to do it...

Let's begin:

1. Head over to our `ViewContoller.swift` file and make the following amendments:

    ```
    import AVFoundation

    private var previewLayer: AVCaptureVideoPreviewLayer! = nil
    var captureSession = AVCaptureSession()
    var bufferSize: CGSize = .zero
    var rootLayer: CALayer! = nil
    private let videoDataOutput = AVCaptureVideoDataOutput()
    private let videoDataOutputQueue = DispatchQueue(
    ```

```
        label: "video.data.output.queue",
        qos: .userInitiated,
        attributes: [],
        autoreleaseFrequency: .workItem)
```

2. Now, create a function called `setupCaptureSession()`. We'll start by adding the following:

```
func setupCaptureSession() {
    var deviceInput: AVCaptureDeviceInput!
    guard let videoDevice = AVCaptureDevice.DiscoverySession(
      deviceTypes: [.builtInWideAngleCamera],
      mediaType: .video,
      position: .back).devices.first else { return }
    do {
        deviceInput = try AVCaptureDeviceInput(device:
videoDevice)
    } catch {
        print(error.localizedDescription)
        return
    }

    // More code to follow
}
```

In the preceding code, we are checking our device for an available camera, specifically `.builtInWideAngleCamera` at the back. If no device can be found, our guard will fail.

3. Next, we initialize `AVCaptureDeviceInput` with our new `videoDevice` object.

4. Now, continuing in our function, add the following code:

```
captureSession.beginConfiguration()
captureSession.sessionPreset = .medium
guard captureSession.canAddInput(deviceInput) else {
    captureSession.commitConfiguration()
    return
}
captureSession.addInput(deviceInput)
if captureSession.canAddOutput(videoDataOutput) {
    captureSession.addOutput(videoDataOutput)
    videoDataOutput.setSampleBufferDelegate(self,
        queue: videoDataOutputQueue)
} else {
    captureSession.commitConfiguration()
    return
}
```

```
do {
    try videoDevice.lockForConfiguration()
    let dimensions =
CMVideoFormatDescriptionGetDimensions(videoDevice.activeFormat.
formatDescription)
    bufferSize.width = CGFloat(dimensions.width)
    bufferSize.height = CGFloat(dimensions.height)
    videoDevice.unlockForConfiguration()
} catch {
    print(error)
}
```

Essentially, here, we are attaching our device to a capture session, allowing us to stream what the device input (camera) is processing programmatically straight into our code. Now, we just have to point this at our view so that we can see the output.

5. Add the following additional code to our function:

```
captureSession.commitConfiguration()
previewLayer = AVCaptureVideoPreviewLayer(session:
captureSession)
previewLayer.videoGravity = AVLayerVideoGravity.resizeAspectFill
rootLayer = imageView.layer
previewLayer.frame = rootLayer.bounds
rootLayer.addSublayer(previewLayer)
```

With the code we've just added, we are essentially creating a visible layer from our current capture session. In order for us to process this on our screen, we need to assign this to rootLayer (the CALayer variable we added earlier). While this seems a little overkill and we could just add this to the layer of UIImageView, we're prepping for something we need to do in our next recipe.

6. Finally, complete the code in our function with the following:

```
captureSession.startRunning()
```

Now, with our camera and device all set up, it's time to set the camera rolling.

Go ahead and run the app. Note again that this will only work on a real device and not a simulator. All going well, you should have a live stream from your camera.

How it works...

The best way to explain this would be to think of the capture session as a wrapper or a configuration between the device's hardware and software. The camera hardware has a lot of options, so we configure our capture session to pick out what we want for our particular instance.

Let's look back at this line of code:

```
AVCaptureDevice.DiscoverySession(deviceTypes:
[.builtInWideAngleCamera], mediaType: .video, position: .back)
```

Here, you could control the `enum` based on a UI toggle, allowing the user to specify which camera to use. You could even add a way for the user to stop the session by using the following:

```
captureSession.stopRunning()
```

Once the session has been configured, you can then start the session by using the following:

```
captureSession.startRunning()
```

Essentially (albeit at a much more complex level), this is what happens when you switch from the front to the back camera when taking a photo.

With the session captured, we can now stream the output directly to any view we like, just like we did here:

```
previewLayer = AVCaptureVideoPreviewLayer(session: captureSession)
```

The fun comes when we want to manipulate the image that is being streamed, by capturing them one frame at a time. We do this by implementing the `AVCaptureVideoDataOutputSampleBufferDelegate` protocol, which allows us to override the following delegate methods:

```
func captureOutput(_ output: AVCaptureOutput, didOutput sampleBuffer:
CMSampleBuffer, from connection: AVCaptureConnection) { }
```

Notice something familiar here... we're being given `sampleBuffer`, just like we got in `UIImagePickerDelegate`. The difference here is that this will be called with every frame, not just when one is selected.

There's more...

Playing around with capture sessions and `AVCaptureOutput` is an expensive operation. Always make sure you stop your session from running when it's not needed, and make sure your delegates are not unnecessarily processing data when they don't need to be.

Another thing to note is that the initialization of a capture device can be slow in some instances, so make sure you have the appropriate UI to handle the potential blocking it may cause.

Final note: if you are struggling with memory leaks and high CPU times, take a look at a suite of tools called **Instruments**. The Xcode Instruments bundle can offer a wide range of performance tracing tools that can really help you get the most out of your Swift code.

See also

For more information, refer to the following links:

- **Instruments overview**: https://help.apple.com/instruments/mac/current/#/dev7b09c84f5

- **AVFoundation**: https://developer.apple.com/documentation/avfoundation

Using CoreML and the Vision framework to detect objects in real time

We've seen what CoreML can do in terms of object detection, but taking everything we've done so far into account, we can certainly go a step further. Apple's Vision framework offers a unique set of detection tools from landmark detection and face detection in images to tracking recognition.

With the latter, tracking recognition, the Vision framework allows us to take models built with CoreML and use them in conjunction with CoreML's object detection to identify and track the object in question.

In this section, we'll take everything we've learned so far, from how AVFoundation works to implementing CoreML, and build a real-time object detection app using a device camera.

Getting ready

For this section, you'll need the latest version of Xcode available from the Mac App Store.

Next, head on over to the Apple Developer portal at the following address: https://developer.apple.com/machine-learning/models/.

Here, you will find out a little bit more about the models available for us to download and use in our Xcode project. You'll notice there are options for image models or text models. For this recipe, we're going to be using image models, specifically one called **YOLOv3**, which uses a residual neural network that attempts to identify and classify what it perceives to be the dominant object in the image.

> **Note**
>
> For more information on the different types of machine learning models, see the links in the *See also* section at the end of this recipe.

From here, download the **YOLOv3.mlmodel** (32-bit) model.

Once downloaded, add this to your Xcode project by simply dragging it into the File Explorer tree in our previous app.

How to do it...

We'll start by creating a new `UIViewController` for all our vision work, in Xcode:

1. Go to **File | New | File**.
2. Choose **Cocoa Touch Class**.
3. Name this `VisionViewController`.
4. Make this a subclass of `UIViewController`.

With that done, we can now head on over to our new `VisionViewController` and add the following code. We'll start by importing the `Vision` framework:

```
import Vision
```

Now, we'll subclass our existing `ViewController` so that we can get the best of both worlds (without the need for copious amounts of code duplication):

```
class VisionViewController: ViewController
```

With that done, we can now override some of our functions in `ViewContoller.swift`. We'll start with `setupCaptureSession()`:

```
override func setupCaptureSession() {
    super.setupCaptureSession()
    setupDetectionLayer()
    updateDetectionLayerGeometry()
    startVision()
}
```

> **Note**
>
> When overriding from another class, always remember to call the base function first. In the case of the preceding code, this can be done by calling `super.setupCaptureSession()`, as used in the preceding code block.

You'll notice some functions in the `VisionViewControler.swift` file that we've not yet created. Let's go through these now one by one:

1. First, we'll add a detection layer to the `rootLayer` that we created earlier. This new `CALayer` will be used as the drawing plane for our detected object area:

    ```
    func setupDetectionLayer() {
        detectionLayer = CALayer()
        detectionLayer.name = "detection.overlay"
        detectionLayer.bounds = CGRect(x: 0.0, y: 0.0, width: buff-
    ```

```
erSize.width, height: bufferSize.height)
    detectionLayer.position = CGPoint(x: rootLayer.bounds.midX,
y: rootLayer.bounds.midY)
    rootLayer.addSublayer(detectionLayer)
}
```

As you can see from the code, we create bounds based on the height and width taken from our bufferSize property (which is being shared back over in our ViewController class).

2. Next, we need to add some geometry to detectionLayer(). This will re-adjust and scale the detection layer based on the device's current geometry:

```
func updateDetectionLayerGeometry() {
    let bounds = rootLayer.bounds
    var scale: CGFloat
    let xScale: CGFloat = bounds.size.width / bufferSize.height
    let yScale: CGFloat = bounds.size.height / bufferSize.width
    scale = max(xScale, yScale)

    if scale.isInfinite {
        scale = 1.0
    }

    CATransaction.begin()
    CATransaction.setValue(kCFBooleanTrue, forKey: kCATransac-
tionDisableActions)

    detectionLayer.setAffineTransform(CGAffineTransform(rotatio-
nAngle: CGFloat(.pi / 2.0)).scaledBy(x: scale, y: -scale))

    detectionLayer.position = CGPoint(x: bounds.midX, y: bounds.
midY)

    CATransaction.commit()
}
```

3. Finally, let's hook up our startVision() function:

```
func startVision(){
    guard let localModel = Bundle.main.url(forResource:
"YOLOv3", withExtension: "mlmodelc") else {
        return
    }

    do {
```

```
        let visionModel = try VNCoreMLModel(for:
MLModel(contentsOf: localModel))
        let objectRecognition = VNCoreMLRequest(model:
visionModel, completionHandler: { (request, error) in
            DispatchQueue.main.async(execute: {
                if let results = request.results {
                    self.visionResults(results)
                }
            })
        })
        self.requests = [objectRecognition]
    } catch let error {
        print(error.localizedDescription)
    }
}
```

4. With this comes a new function, `visionResults()`. Go ahead and create this function in `VisionViewController`, too.

Note

We could have simply used an extension in our original `ViewController` to house all these new functions, but we'd run the risk of overloading our view controller to the point where it could become too unmaintainable.

Also, our logic and extension for `UIImagePicker` were in here, so the separation is nice.

---

5. With this, let's build the `visionResults()` function. We'll do this one section at a time so it all makes sense:

```
func visionResults(_ results: [Any]) {
    CATransaction.begin()
    CATransaction.setValue(kCFBooleanTrue, forKey:
kCATransactionDisableActions)

    detectionLayer?.sublayers = nil

    // More code to follow
}
```

We start with some basic housekeeping; performing `CATransaction` locks in memory any changes we're going to make to `CALayer` before we finally commit them for use. In this code, we'll be modifying `detectionLayer`.

6. Next, we'll iterate around our `results` parameter to pull out anything that is of the `VNRecognizedObjectObservation` class type:

```
for observation in results where observation is
VNRecognizedObjectObservation {
    guard let objectObservation = observation as?
VNRecognizedObjectObservation else {
        continue
    }

    let labelObservation = objectObservation.labels.first
    let objectBounds =
VNImageRectForNormalizedRect(objectObservation.boundingBox,
Int(bufferSize.width), Int(bufferSize.height))

    let shapeLayer = createRoundedRectLayer(with: objectBounds)
    let textLayer = createTextSubLayer(with: objectBounds,
identifier: labelObservation?.identifier ?? "", confidence:
labelObservation?.confidence ?? 0.0)

    shapeLayer.addSublayer(textLayer)
    detectionLayer?.addSublayer(shapeLayer)

    updateDetectionLayerGeometry()
    CATransaction.commit()
}
```

From this, we'll continue to use Vision to obtain `Rect` and position of the identified object(s) using `VNImageRectForNormalizedRect`. We can also grab some text information about the objects detected and use that too.

7. Finally, we'll gracefully close off any changes to `detectionLayer` and update the geometry to match the detected objects. You'll notice there are two new functions we've just introduced:

```
createRoundedRectLayer()
createTextSubLayer()
```

8. These again are helper functions, one to draw the rectangle of the detected object and the other to write the text. These functions are generic boilerplate sample code that can be obtained from Apple's documentation and can be found in the sample project on GitHub; just copy them into your project (either in `VisionViewController` or your own helper file). Feel free to play around with these to suit your needs.

9. One thing I will mention: you'll notice how we do all this again using `layers` rather than adding `UIView` and `UILabel`. This again is because `UIKit` is a wrapper around a lot of core functionality. But adding a `UIKit` component on top of another component is unnecessary, and with what is already an intense program, this could be performed much more efficiently by updating and manipulating the layers directly on a `UIKit` object.

With our `AVFoundation` camera streaming in place and Vision and CoreML ready to do their magic, there is one final override we need to add to `VisionViewController`:

```
override func captureOutput(_ output: AVCaptureOutput, didOutput
sampleBuffer: CMSampleBuffer, from connection: AVCaptureConnection) {
    guard let pixelBuffer = CMSampleBufferGetImageBuffer(sampleBuffer)
else {
        return
    }

    let exifOrientation = ImageHelper.
exifOrientationFromDeviceOrientation()

    let imageRequestHandler = VNImageRequestHandler(cvPixelBuffer:
pixelBuffer, orientation: exifOrientation, options: [:])

    do {
        try imageRequestHandler.perform(self.requests)
    } catch {
        print(error)
    }
}
```

Using the delegate for `AVFoundation`, we grab each frame again, converting this to `CVPixelBuffer` in order to create `VNImageRequestHander`. This now kicks off the requests in our `startVision()` function, stitching everything together nicely.

We're almost done; let's finish off with some bits and pieces to tie all this together now:

1. Head on over to `ViewController.swift` and add the following function, using the `IBAction` keyword, then add the logic from `UISegmentedControl`, which we created earlier:

```
@IBAction func onInputTypeSelected(_ sender: UISegmentedControl)
{
    switch sender.selectedSegmentIndex {
    case 0:
        captureSession.stopRunning()
    case 1:
        startLivePreview()
    default:
        print("Default case")
    }
}
```

2. Now, create a function called `startLivePreview()`:

```
func startLivePreview() {
    captureSession.startRunning()
}
```

3. Remove `captureSession.startRunning()` from `setupCaptureSession()`.

4. Finally, in our `Main.storyboard` view controller, change the class from `ViewController` to `VisionViewController`.

5. Now, go ahead and run the app. All going well, you should be live-detecting images with an overlay that looks like this:

Figure 10.2 – Vision detection

As you can see, both Vision and CoreML have successfully detected my cell phone and its location in the image (all in real time).

How it works...

A high-level overview goes something like this:

1. Capture a real-time camera feed (using `AVFoundation`).

2. Use a trained CoreML model to detect whether the image contains an object (that it recognizes).

3. Use Vision to detect the position of the object in the picture.

We covered the camera streaming elements in the previous recipe, but let's take a deeper look at how *step 2* and *step 3* work.

Let's actually start with *step 3*. We saw in the last section how we use `VNImageRequestHander` to pass back `CVPixelBuffer` of each image frame. This now fires off calls in our `startVision()` function, so let's take a closer look at this.

First, we grab our model from the app's bundle, so that we can pass this over to Vision later on to prepare the object recognition:

```
guard let localModel = Bundle.main.url(forResource: "YOLOv3",
withExtension: "mlmodelc") else {
    return
}
```

Next, we head back to *step 2*, where we create an instance of `VNCoreMLModel()`, passing in the `localModel` that we just created.

```
let visionModel = try VNCoreMLModel(for: MLModel(contentsOf:
localModel))
```

With the newly created `visionModel`, which is our `VNCoreMLModel`, we can now create our `VNCoreMLRequest` call, along with its completion handler, which will fire from requests that come in via our `AVFoundation` delegate.

This one simple request does the work of both the Vision framework and CoreML – first detecting whether an object is found, then supplying us with the details on where that object is located inside the image.

This is where the bulk of our work is done. If you look again at our `visionResults()` function and all the helper functions within, these are merely ways of parsing data that has come back, and, in turn, decorating our view.

In our *results* from the `VNCoreMLRequest()` response, we take an instance of `VNRecognize-dObjectObservation`, which, in turn, gives us two properties: a label (of what CoreML thinks it has found) along with a confidence score.

See also

For more information on `CALayer`, refer to `https://developer.apple.com/documentation/quartzcore/calayer`.

Immersive Swift with ARKit and Augmented Reality

Imagine being able to visualize data, images, 3D objects, or effects in tandem with the world around you. It would be short of those experiences happening right before you, but when blended in with your environment in real time, it can seem like it is happening before your eyes.

This is **Augmented Reality** (**AR**). It cannot alter reality itself. Nor does it try to replace it completely and artificially like virtual reality. It simply serves to augment, or enhance, reality by building upon it.

With the computing power, cameras, and sensors we carry around with us in our devices, it's no wonder that AR is such a versatile and awesome feature. This is where **ARKit**, an Apple framework that allows us to easily and quickly build AR experiences, comes in.

By the end of this chapter, you will know how to use ARKit to create simple yet immersive AR experiences for your apps.

In this chapter, we will cover the following recipes:

- Surface detection with ARKit
- Using 3D models with ARKit
- Using Reality Composer Pro for visionOS

Technical requirements

You can find the code files present in this chapter on GitHub at `https://github.com/PacktPublishing/Swift-Cookbook-Third-Edition/tree/main/Chapter%2011`.

Surface detection with ARKit

In this first recipe, we're going to create a simple ARKit app that detects horizontal surfaces (or planes). Then, having that detection in place, we'll use that information to display a 3D object in our AR world!

We'll also do this in SwiftUI with **ARView,** which builds a view for our AR world to be displayed through. ARView is primarily a UIKit component. But with a little help from **UIViewRepresentable**, we can still take advantage of SwiftUI's goodness (and with less code overall).

Getting ready

For this recipe, to run the app, you will need a physical device that is compatible with ARKit, as you cannot test these apps in the simulator – due to needing a camera. Generally, any device that is running on iOS 11 or later will be okay.

How to do it...

With Xcode open, let's get started:

1. Create a new project in Xcode. Go to **File | New | Project | iOS App**. For this app, be sure to choose SwiftUI for the interface.

2. We will need to declare a few attributes for our app in our app's `Info.plist`. Simply select our project in **Navigator**, select the project target, and go to **Info**. You'll need to add **Privacy**, **Camera Usage Description**, and **Required Device Capabilities** (`Item 0 = ARKit`).

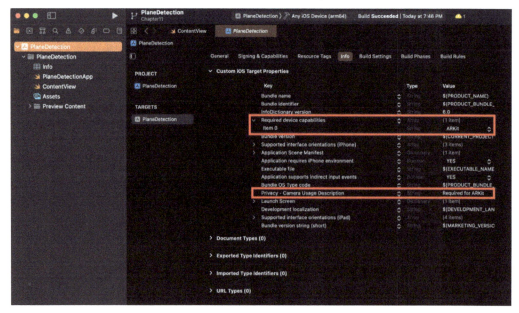

Figure 11.1 – Adding the two attributes to the app under the Info tab

3. In `ContentView.swift`, we'll start by importing `ARKit` and `RealityKit`:

    ```
    import ARKit
    import RealityKit
    ```

4. Next, we'll create a struct conforming to `UIViewRepresentable` that will wrap an ARView:

    ```
    struct ARViewContainer: UIViewRepresentable {
        func makeUIView(context: Context) -> ARView {
            let arView = ARView(frame: .zero)
            // More code to follow
            return arView
        }


        func updateUIView(_ uiView: ARView, context: Context) {
            // No need to update the view in this example
        }
    }
    ```

5. Before we return our `ARView`, we'll add some configuration to it that explicitly states that we want to track our device against objects in our environment, specifically looking for horizontal planes (such as table tops, floors, etc.):

    ```
    let configuration = ARWorldTrackingConfiguration()
    configuration.planeDetection = [.horizontal]
    arView.session.run(configuration)
    ```

6. To be able to see that we're detecting planes, let's add an object to our `ARView`. First, we'll create an entity in the form of `AnchorEntity` for our object so it knows where to place itself:

    ```
    let anchorEntity = AnchorEntity(.plane(.horizontal,
    classification: .any, minimumBounds: [0.2, 0.2]))
    ```

7. Then, we'll create a basic red box, taking advantage of `ModelEntity` to whip one up in just a few lines of Swift, and add it to our `AnchorEntity`:

    ```
    let box = MeshResource.generateBox(width: 0.1, height: 0.1,
    depth: 0.1)
    let boxMaterial = SimpleMaterial(color: .red, isMetallic: false)
    let boxModel = ModelEntity(mesh: box, materials: [boxMaterial])
    anchorEntity.addChild(boxModel)
    ```

8. Next, add our anchor to our `ARView`:

    ```
    arView.scene.anchors.append(anchorEntity)
    ```

9. Lastly, swap out the body of our `ContentView` to display our `ARViewContainer`:

```
struct ContentView: View {
    var body: some View {
        ARViewContainer()
    }
}
```

10. Now, with your device plugged in, run the app. The first time the app runs, we'll need to give permission to use the device's camera. Once authorized, move your camera around and watch a red box appear on any surface nearby:

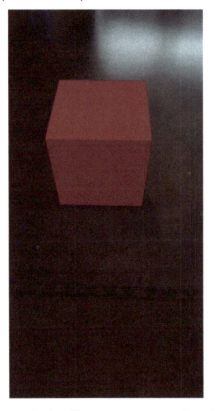

Figure 11.2 – A red box appears on my coffee table!

How it works...

It may be surprising how little code it takes to get an `ARView` up and running. And that's even including wrapping it in a `UIViewRepresentable`. To summarize, `UIViewRepresentable` is a protocol that wraps a UIKit-based view and, literally, represents it to SwiftUI. For more information, check out the link in the *See also* section.

Once we set our `ARView` and return it, we still have to start an AR session. To do that, we need to set a configuration first. That's where we created `ARWorldTrackingConfiguration`, telling our AR that we want to define the relationship between where our device is located and the objects in the environment around it. We then specify that we want to find horizontal planes in our environment by setting `planeDetection`. Lastly, with our configuration set, we pass it in as we tell our `ARView` to begin running its session:

```
let configuration = ARWorldTrackingConfiguration()
configuration.planeDetection = [.horizontal]
arView.session.run(configuration)
```

Next, we create an `AnchorEntity` that will tap into any horizontal plane detected during our session. This will be used for any objects that we want to display in our AR world to decide where it's located relative to reality:

```
let anchorEntity = AnchorEntity(.plane(.horizontal, classification:
.any, minimumBounds: [0.2, 0.2]))
```

Then, we create a simple red box. For this, we take advantage of another framework, called **RealityKit**, which provides an API that works together with `ARKit` for the purposes of simulating and rendering 3D objects. In three lines of Swift, we create a box (`MeshResource.generateBox`), define its surface (`SimpleMaterial`), and put it together as a prepared `ModelEntity`, ready to be inserted in our `ARView`:

```
let box = MeshResource.generateBox(width: 0.1, height: 0.1, depth:
0.1)
let boxMaterial = SimpleMaterial(color: .red, isMetallic: false)
let boxModel = ModelEntity(mesh: box, materials: [boxMaterial])
```

We then attach our new object to our `AnchorEntity` so it now knows where to rest in our AR world:

```
anchorEntity.addChild(boxModel)
```

Lastly, we attach our `AnchorEntity` to our `ARView`:

```
arView.scene.anchors.append(anchorEntity)
```

Once we declare our `ARViewContainer` in `ContentView` and run our app on a device, we see our red box appear on the nearest horizontal surface in all its glory!

There's more...

One thing about this implementation is, if someone were to walk in front of the camera, our box would still show up in front of the person – even if, distance-wise, the person were standing closer to the camera than the box.

We can account for this difference with a simple addition to the configuration!

```
configuration.frameSemantics.insert(.personSegmentationWithDepth)
```

`ARKit` provides the option to include **person occlusion** in the configuration of `ARView`. This means that it will detect a person's presence (could be a whole body or just part) and calculate the distance and location relative to the AR world. It will then compare it to any AR objects, determine which is closer to the camera, and decide what should be shown. In *Figure 11.3*, you'll see that my hand is behind the red box.

Figure 11.3 – Now my hand is behind!

In *Figure 11.4*, `ARKit` determines my hand is in front of the box. Therefore, it displays my hand while still showing the box, though as if it were behind my hand. All of this, mind you, in real time!

Figure 11.4 – And now that I'm closer to the camera, the box is behind!

See also

- For more information on `ARKit`, refer to `https://developer.apple.com/documentation/arkit`

- For more information on `RealityKit`, refer to `https://developer.apple.com/documentation/realitykit`

- For more information on `UIViewRepresentable`, refer to `https://developer.apple.com/documentation/swiftui/uiviewrepresentable`

- For more information on people occlusion, refer to `https://developer.apple.com/documentation/arkit/arkit_in_ios/camera_lighting_and_effects/occluding_virtual_content_with_people`

Using 3D models with ARKit

In the previous recipe, we looked at how to detect horizontal planes in the environment, and how to place a 3D object in the AR world. In this recipe, we are going to build on our existing knowledge and use a more realistic object that we might find in the real world – something that is more fun and recognizable. This is where AR really starts to come to life.

Getting ready

While this recipe builds on top of the previous recipe, it is recommended in this recipe that we create a new Xcode project. This is because Xcode provides us with a more efficient way of setting up an AR app that will look just like our previous recipe.

How to do it...

Let's get started with the new recipe:

1. Create a new project in Xcode. Go to **File | New | Project | Augmented Reality App**. For this app, be sure to choose **SwiftUI** for **Interface** and **RealityKit** for **Content Technology**:

Figure 11.5 – Selecting Augmented Reality App

Note that we're not using **App**, but rather **Augmented Reality App**, which comes with AR-specific boilerplate already in place.

2. In `ContentView`, you'll notice a lot looks similar to our previous recipe, especially `ARViewContainer`. We're going to alter this code later on.

3. Next, we need to find a 3D model that we can use in our new AR app. To save us some time, we can download one of these from Apple's AR Quick Look Gallery: `https://developer.apple.com/augmented-reality/quick-look/`.

4. For now, let's use `Toy Biplane`, found under the **3D model** section of the AR Quick Look Gallery. Once the 3D model has been downloaded, we can then open up the file to see the preview and look around.

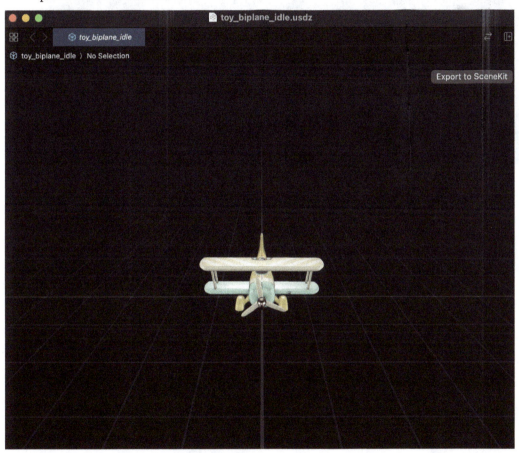

Figure 11.6 – Preview of the downloaded 3D model from the AR Quick Look Gallery

5. We can then drop the 3D model file into our new **Augmented Reality App** project. Make sure you select the option to copy the items if needed:

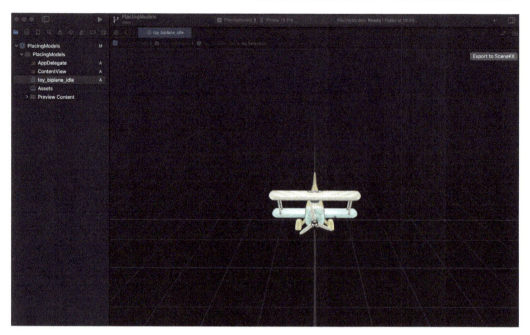

Figure 11.7 – The downloaded 3D model added to the Xcode project

6. Now, time for some code. Head on over to our `ARViewContainer` in `ContentView` and replace any existing code with the following, to load our 3D model and attach it to our `ARView` scene:

```
struct ARViewContainer: UIViewRepresentable {
    func makeUIView(context: Context) -> ARView {
        let arView = ARView(frame: .zero)
        let planeEntity = try! ModelEntity.loadModel(named:
"toy_biplane_idle")
        let anchorEntity = AnchorEntity(.plane(.horizontal,
classification: .any, minimumBounds: [0.2,0.2]))
        anchorEntity.addChild(planeEntity)
        arView.scene.anchors.append(anchorEntity)
        return arView
    }

    func updateUIView(_ uiView: ARView, context: Context) {
        // No need to update the view in this example
    }
}
```

7. Now, with your device plugged in, run the app. The first time the app runs, we'll need to give permission to use the device's camera. Once authorized, move your camera around and watch the 3D model appear on any surface nearby:

Figure 11.8 – The toy biplane appears on my desk!

How it works...

In not much additional time, we have swapped out our basic red box for a realistic and interactive 3D model. In this recipe, we used Toy Biplane, and now you have the knowledge to repeat this again for any other 3D models you can find. Remember, you can always refer back to Apple's AR Quick Look Gallery.

To recap, we created a new AR app, downloaded a 3D model, imported it into our Xcode project, and updated a few lines of code.

Firstly, we created our ARView, as we have done before:

```
let arView = ARView(frame: .zero)
```

Next, we loaded Toy Biplane as ModelEntity, which prepared our 3D model for the AR scene, using the exact filename for the 3D model that can be seen in our project navigator:

```
let planeEntity = try! ModelEntity.loadModel(named: "toy_biplane_
idle")
```

Then, we created an `AnchorEntity` so that our `ModelEntity` can place itself in the AR world – specifically on a horizontal plane:

```
let anchorEntity = AnchorEntity(.plane(.horizontal, classification:
.any, minimumBounds: [0.2,0.2]))
anchorEntity.addChild(planeEntity)
```

Finally, we can attach `AnchorEntity` to the `ARView` scene, before returning the `ARView`:

```
arView.scene.anchors.append(anchorEntity)
return arView
```

That's all we needed to do to bring our AR world to life! Now, when we run the app, we will see the 3D models in action!

There's more...

There is one very special feature about the `Toy Biplane` 3D model that we haven't looked at yet, which is that this particular 3D model has some animations – how exciting! You may have noticed that the 3D model file has a **USDZ** format – which stands for **Universal Scene Description Zipped**. With this being a scene file, animations can be described for the 3D object as part of the scene, which can be executed when triggered correctly. By default, we do not see these animations after simply loading `ModelEntity`.

We can add one additional snippet of code before we return our `ARView` in the `makeUIView` function, to start seeing the animations in our AR world:

```
planeEntity.availableAnimations.forEach {
    planeEntity.playAnimation($0.repeat)
}
```

Now, once we run the app again, we'll be able to see that the toy biplane is now hovering on the spot – with its front propeller spinning around!

Using Reality Composer Pro for visionOS

As we saw in the last recipe, we can create 3D objects and display them in AR quite simply in Swift. Another way in which we can create AR scenes is by using a GUI tool that automatically generates an AR scene that we can use directly in our app.

Enter **Reality Composer Pro**, an Apple tool that does exactly that for the visionOS platform. For more information, check out the link in the *See also* section.

Getting ready

For this recipe, we will be using Reality Composer Pro, which is available with Xcode 15. This recipe will focus more on creating an AR experience with the visionOS SDK, as there is currently a limitation on using Reality Composer Pro with the standard SDKs (such as iOS).

How to do it...

Let's get started with this recipe:

1. Create a new project in Xcode. Go to **File** | **New** | **Project** | **VisionOS** | **App**. For this app, be sure to choose **Volume** for **Initial Scene** and **RealityKit** for **Immersive Space Renderer**:

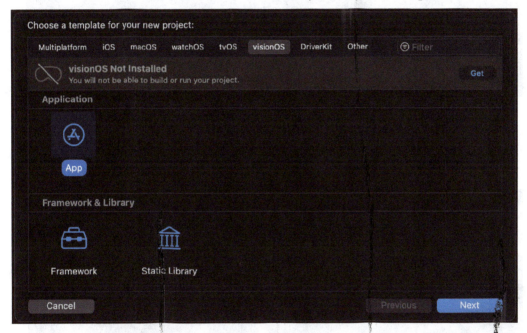

Figure 11.9 – Selecting visionOS | App

Note that on this occasion, we are building a visionOS app, instead of an iOS app, which comes with its own set of AR-specific boilerplate already in place. You may need to install the visionOS simulator in order to run the app later on.

2. In `ContentView`, you'll notice a whole lot of new code that you have not seen before. The only thing you need to be aware of right now is that we have an Entity, called `Scene`, that is being loaded into a RealityView.

3. Next, we can open up our `Scene`, which will be a Reality Composer Pro project. This can be found inside **Packages** | **RealityKitContent** | **Sources** | **RealityKitContent** | **RealityKitContent**.

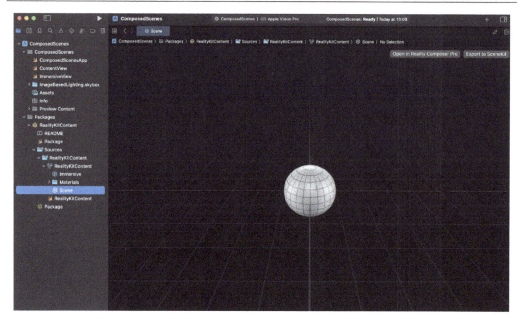

Figure 11.10 – The Reality Composer Pro previewer in Xcode with
a button to open in the Reality Composer Pro app

4.   Go ahead and open up the scene in the Reality Composer Pro app, from the previewer in Xcode:

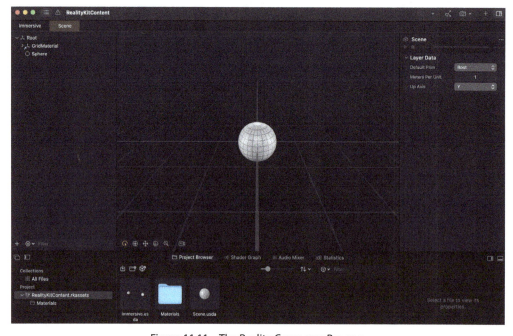

Figure 11.11 – The Reality Composer Pro app

5. From the Reality Composer Pro app, we can now see our AR scene with a number of tools and configurations available to us.

6. On the toolbar, hit +. Here, you'll find all sorts of pre-made 3D objects available to use. For now, search `Plane` and select the object listed as `Toy Biplane`:

Figure 11.12 – Choosing the Toy Biplane object to add to our scene

7. Once the `Toy Biplane` model has been added, click the *Sphere* in the component navigator, on the left, then hit *Delete*. We should be left with just the `Toy Biplane` model in our scene.

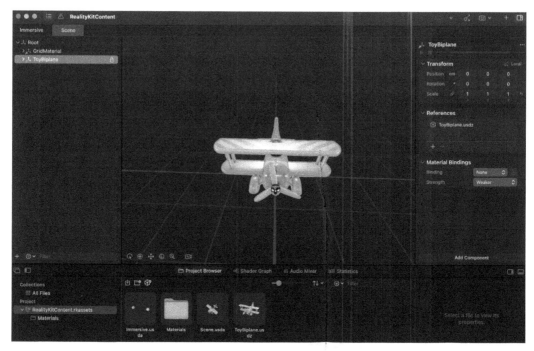

Figure 11.13 – Replacing the sphere with the toy biplane in our scene

8. Save and go back to our project in Xcode. You may notice that the structure of the **RealityKitContent** folder has changed slightly, with the scene now updated with our Toy Biplane.

9. We don't need to add any additional code anywhere, as we are already loading our Scene in RealityView.

10. Finally, with the visionOS simulator installed, hit **Run** and see how the 3D model of Toy Biplane is added to the virtual room:

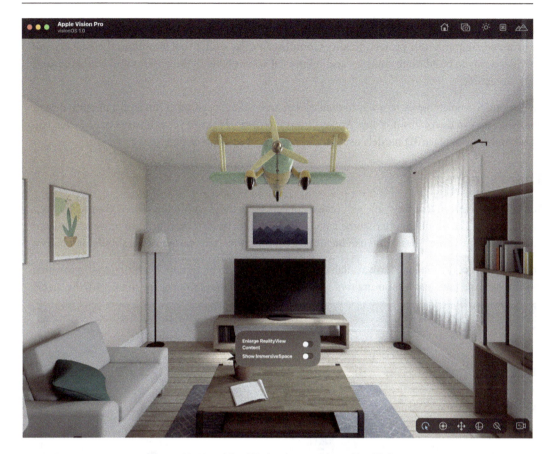

Figure 11.14 – visionOS simulator with our Toy Biplane

How it works...

The first thing to notice is that choosing a visionOS app afforded us several boilerplate items, all still using SwiftUI. This included the **RealityKitContent** package with a basic Reality Composer Pro project already in place, along with the code needed to load our Scene into RealityView. We were then able to add our own 3D model to the scene, all using Reality Composer Pro.

While we didn't get to mess around much with Swift in this recipe, there are some key things to point out. The main thing is how much code is already in place for us. We don't require any code to determine plane detection, or to set an anchor, as we did with the iOS app. The nature of visionOS means that the system is able to handle all of the detection and placement for us.

RealityKitContent provided by Reality Composer Pro bundles up our scene into a package that can be dropped right into our project. Reality Composer Pro as a GUI tool simplifies the process and provides a way to build more complex and robust AR scenes with all the options and configurations easily accessible.

The visionOS simulator gives us a way to preview our app without needing to run it on a physical device, much like we can do for any iOS app. Apple Vision Pro devices are not yet mainstream; however, you now know how to place 3D models using the visionOS SDK ahead of time.

There's more...

You will have noticed that the scene we created appears quite small within the visionOS simulator as it first launches. The simulator offers some navigation tools that make it easy for us to move around and interact with the scene. These tools can be found in the bottom-right corner of the simulator window.

To name a few of the tools we have available to us, there is an interaction tool, a tilting tool, a panning tool, an orbiting tool, and a zooming tool. Using these tools allows us to engage more with the scene.

Figure 11.15 – Using the navigation tools within the VisionOS simulator

Additionally, you'll have noticed two options available to us in the scene: **Enlarge RealityView Content** and **Show ImmersiveSpace**. We can enlarge the content so that it appears more clearly to us, or we can choose to merge our `Scene` into another immersive scene, which is also available to adjust within the same Reality Composer Pro project.

See also

- For more information on AR tools, refer to `https://developer.apple.com/augmented-reality/tools/`

- For more information on Reality Composer Pro, refer to `https://developer.apple.com/wwdc23/10083`

- For more information on Vision Pro, refer to `https://www.apple.com/apple-vision-pro/`

12
Visualizing Data with Swift Charts

An important way for information and data to be presented more simply and effectively is by conveying them in a more visually understandable way. Looking at a list of data such as a spreadsheet can be difficult for a majority of users, especially the more robust and complex it becomes.

A way to make data more readable and understandable is through charts. They transform information in a way that's more aesthetically pleasing while providing opportunities to better represent data within data (e.g., patterns, trends, etc.). Charts have helped visualize data in software for decades, even before graphics were a thing and they needed to be built using text-based characters.

Charts are incredibly useful in apps today, as well. That's why Apple introduced **Swift Charts**, a new framework that offers to centralize and take on much of the heavy lifting when it comes to building charts.

By the end of this chapter, you will learn how to leverage Swift Charts to build informative and even interactive charts so you can better inform and educate your app users.

In this chapter, we will cover the following recipes:

- Building a chart with data
- Displaying multiple datasets
- Exploring chart marks and modifiers

Technical requirements

You can find the code files present in this chapter on GitHub at https://github.com/PacktPublishing/Swift-Cookbook-Third-Edition/tree/main/Chapter%2012.

Building a chart with data

In this first recipe, we're going to create a simple Swift Chart that will help us visualize work performance based on the cups of coffee they've had (scientific accuracy questionable). We'll do this in a barebones SwiftUI app.

Getting ready

For all the recipes in this chapter, you'll need the latest version of Xcode available from the Mac App Store.

How to do it...

With Xcode open, let's get started:

1. Create a new project in Xcode. Go to **File** | **New** | **Project** | **iOS App**. Be sure to choose **SwiftUI** for the interface.

2. In our project, select **New File** | **SwiftUI View**, and name this file `CoffeePerformance`.

3. In our new file, the first thing is to add Swift Charts by adding `import Charts` to the top of our file.

4. Let's create some data for our chart to consume. First, make a new struct called `PerformanceInfo`:

    ```
    struct PerformanceInfo: Identifiable {
        var cups: Int
        var rating: Int
        var id = UUID()
    }
    ```

5. Using `struct`, we'll make a collection of data points:

    ```
    var dannyPerfInfo: [PerformanceInfo] = [
        .init(cups: 0, rating: 1),
        .init(cups: 1, rating: 2),
        .init(cups: 2, rating: 4),
        .init(cups: 3, rating: 5),
        .init(cups: 4, rating: 3),
        .init(cups: 5, rating: 1),
        .init(cups: 6, rating: 0)
    ]
    ```

6. In body of `CoffeePerformance`, set up `VStack` with `Text` inside:

    ```
    VStack (alignment: .leading) {
        Text("Danny's Coffee ☕")
    ```

```
        }
        .padding()
```

7. Below `Text`, let's add `Chart` and provide the data we set up earlier. Then, `BarMark` can be added inside the `Chart`, as follows:

```
Chart(dannyPerfInfo) { perfInfo in
    BarMark(
        x: .value("Cups of Coffee", perfInfo.cups),
        y: .value("Rating", perfInfo.rating)
    )
}
```

We'll immediately begin to see the bar chart in our Previews, as shown in *Figure 12.1*:

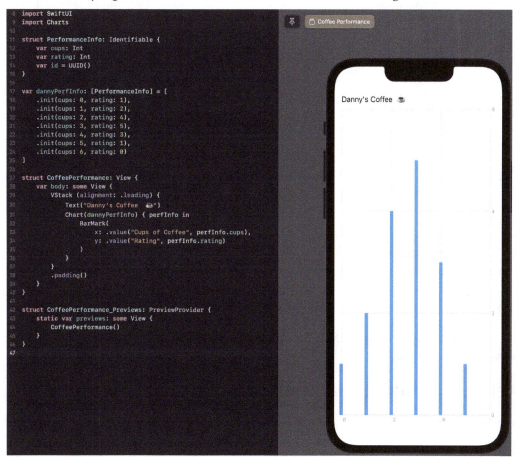

Figure 12.1 – A bar chart displaying Coffee Performance information in Xcode Previews

How it works...

Anyone who's worked with charts in code before has probably been taken aback that it takes just six lines of code to display data in a bar chart. What is even more amazing is that we can immediately see the results in Xcode Previews. Let's break down the small amount of code we've done so far.

We started by constructing our struct, `PerformanceInfo`, which simply defined a single point of data. In our case, we added two variables capturing a performance rating reached at a specific number of cups of coffee. For our chart to be able to iterate over our data, we also needed to make sure we conformed to `Identifiable`, requiring us to also add `id` as a variable.

We then created a simple array of `PerformanceInfo` that held our *scientifically-accurate* data.

Jumping to `Chart`, we simply passed in our data (`dannyPerfInfo`) as a parameter into a new `Chart`. Like `ForEach`, `Chart` will then provide us with each point of data in the collection we pass in as `perfInfo`:

```
Chart(dannyPerfInfo) { perfInfo in
```

We then define the type of chart we want to display by declaring a type of `Mark`. To create a bar chart, we used `BarMark` (we'll look into different types of marks in the *Exploring chart marks and modifiers* recipe). For a `BarMark` to know its own value, we feed it both an `x` and `y` value, based on the data we've already defined and fed into our chart:

```
BarMark(
    x: .value("Cups of Coffee", perfInfo.cups),
    y: .value("Rating", perfInfo.rating)
)
```

This is all it takes to get started on building a comprehensive chart using Swift Charts. While this may seem basic, it doesn't take much to customize and build on this chart to display even more data in an understandable and visually pleasing way. We'll dive deeper into this in the next few recipes.

See also

For more information on Swift Charts, refer to `https://developer.apple.com/documentation/charts`.

Displaying multiple datasets

In this recipe, we'll take the chart we just built and enhance it so we can add another set of coffee performance rating data. Charts are great for comparing sets of data because a viewer can better see how the sets align or differ, and by how much.

Because Swift Charts streamlines building charts for all kinds of data and scenarios, it won't take much to add our new set. However, we may want to make a few modifications to make our data stand out clearly.

How to do it...

Let's start where we left off in our previous project:

1. We'll first start by adding another set of coffee performance data:

```
var johnPerfInfo: [PerformanceInfo] = [
    .init(cups: 0, rating: 1),
    .init(cups: 1, rating: 1),
    .init(cups: 2, rating: 1),
    .init(cups: 3, rating: 3),
    .init(cups: 4, rating: 5),
    .init(cups: 5, rating: 2),
    .init(cups: 6, rating: 0)
]
```

2. Let's add our new set and join it with our old set in a tuple array:

```
let multiplePerformers = [
    (name: "Danny", info: dannyPerfInfo),
    (name: "John", info: johnPerfInfo)
]
```

3. Because the structure of our data has changed, we will instead feed `Chart` with `multiplePerformers`:

```
Chart (multiplePerformers, id: \.name) { performer in
```

4. Next, we'll wrap `BarMark` with `ForEach` based on each `perfomer.info`:

```
ForEach(performer.info) { perfInfo in
    BarMark(
        x: .value("Cups of Coffee", perfInfo.cups),
        y: .value("Rating", perfInfo.rating)
    )
}
```

It may be hard to notice, but our chart will now reflect our multiple datasets as visible in *Figure 12.2*:

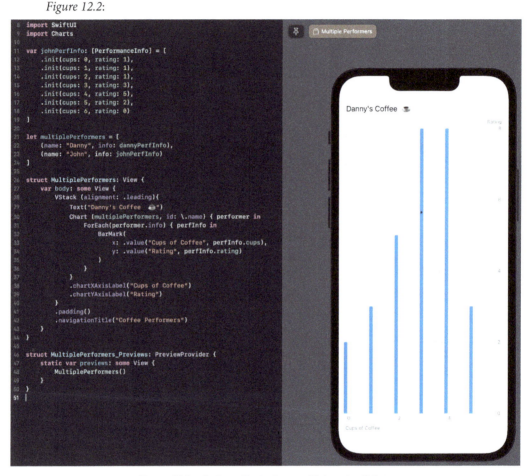

```swift
import SwiftUI
import Charts

var johnPerfInfo: [PerformanceInfo] = [
    .init(cups: 0, rating: 1),
    .init(cups: 1, rating: 1),
    .init(cups: 2, rating: 1),
    .init(cups: 3, rating: 3),
    .init(cups: 4, rating: 5),
    .init(cups: 5, rating: 2),
    .init(cups: 6, rating: 0)
]

let multiplePerformers = [
    (name: "Danny", info: dannyPerfInfo),
    (name: "John", info: johnPerfInfo)
]

struct MultiplePerformers: View {
    var body: some View {
        VStack (alignment: .leading){
            Text("Danny's Coffee ☕")
            Chart (multiplePerformers, id: \.name) { performer in
                ForEach(performer.info) { perfInfo in
                    BarMark(
                        x: .value("Cups of Coffee", perfInfo.cups),
                        y: .value("Rating", perfInfo.rating)
                    )
                }
            }
            .chartXAxisLabel("Cups of Coffee")
            .chartYAxisLabel("Rating")
        }
        .padding()
        .navigationTitle("Coffee Performers")
    }
}

struct MultiplePerformers_Previews: PreviewProvider {
    static var previews: some View {
        MultiplePerformers()
    }
}
```

Figure 12.2 – A chart displaying multiple sets of data

5. We'll add one modifier to `BarMark` to change the foreground color of each set:

```swift
.foregroundStyle(by: .value("Performer", performer.name))
```

This is demonstrated in *Figure 12.3*.

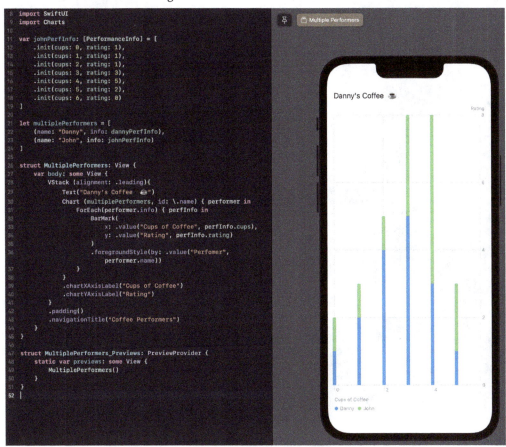

Figure 12.3 – A chart with multiple datasets distinguished by color

6. We'll add another modifier to offset each set's mark, as reflected in *Figure 12.4*:

```
.position(by: .value("Performer", performer.name), axis:
.horizontal, span: .inset(15))
```

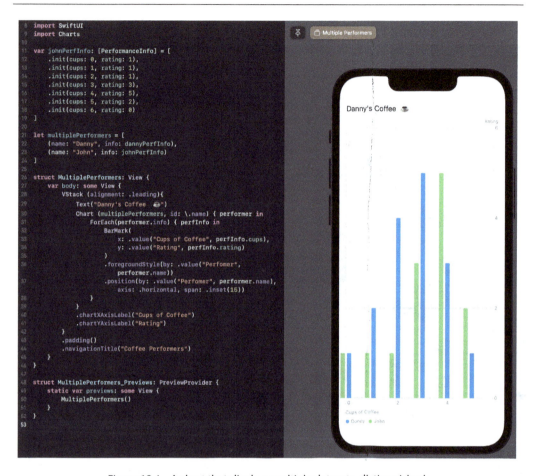

Figure 12.4 – A chart that displays multiple datasets, distinguished
by color, clearly set apart from each other

With all that in place, let's break down the changes we just made to understand what just happened
a little more.

How it works...

First, we created our new dataset (this time for John) and joined it together with our old (Danny's)
dataset. We used an array of tuples (with the name and info named parameters) to avoid having
to define and use a new type of struct. Tuples are great for when we don't need to overelaborate data
but still need to set a relationship between them:

```
let multiplePerformers = [
    (name: "Danny", info: dannyPerfInfo),
    (name: "John", info: johnPerfInfo)
]
```

Charts already know how to take in data as a parameter, as we saw in the previous recipe. However, our data has changed and now comprises a collection of a collection. To adjust for this, we made a few changes so that `Chart` would understand this composition better and display our data appropriately.

First, we pass into `Chart` our highest-level collection, `multiplePerformers`. Because we made a simple tuple array, we also needed to define which value in the tuple for `Chart` to iterate over (in this case, `name` is our identifier).

To then have `Chart` iterate over our lower-level collection, we wrapped `BarMark` in `ForEach`, which iterates over the information of the current performer:

```
ForEach(performer.info) { perfInfo in
    BarMark(
        x: .value("Cups of Coffee", perfInfo.cups),
        y: .value("Rating", perfInfo.rating)
    )
}
```

Notice that in just about everything from `perfInfo` in through our entire `BarMark` implementation, the code stayed the same from our previous recipe. That's because we only needed to define to `Chart` how our new higher-level collection worked in relation to the collection it was already familiar with.

The last thing we did was add two view modifiers that helped `Chart` better distinguish between the two sets of data. It's hard to tell, but without modifications, the sets are stacked on top of each other; this doesn't help anyone reading the chart compare the datasets.

By adding the `foregroundStyle` modifier, we distinguished the sets based on the performer's name. From there, Swift Charts automatically assigns different colors to each dataset and even provides a legend at the bottom of our chart that defines which color is for which set.

Similarly, we added the `position` modifier so that we could unstack the sets and see them side by side. Again, by defining the value to be based on the performer's name, Swift Charts automatically handles assigning span insets by set.

There are a ton of modifiers available just for Swift Charts to alter and better display any data we want to show to our users. We will dive into more of these, as well as different chart marks, in the next recipe.

Exploring chart marks and modifiers

Now that we've accomplished plugging data into Swift Charts, it's time to learn how we can alter how our chart displays our data. The two most significant ways we can alter the look of our chart are through marks and modifiers. When combined, there are countless ways we can alter the look (and feel) of a chart.

How to do it...

We will start where we left off in our previous project:

1. In `CoffeePerformance.swift`, let's add a small enum of the marks we will be exploring:

   ```
   enum ChartMark {
       case bar, line, area, point, rect
   }
   ```

2. In `CoffeePerformance` (the View itself), we'll place a new `State`:

   ```
   @State var selectedChartMark: ChartMark = .bar
   ```

3. We're going to replace `BarMark` and expand it with a switch over all the different types of `Mark`:

   ```
   switch selectedChartMark {
   case .bar:
       BarMark(
           x: .value("Cups of Coffee", perfInfo.cups),
           y: .value("Rating", perfInfo.rating)
       )
       .foregroundStyle(by: .value("Performer", performer.name))
       .position(by: .value("Performer", performer.name), axis:
   .horizontal, span: .inset(15))
   case .line:
       LineMark(
           x: .value("Cups of Coffee", perfInfo.cups),
           y: .value("Rating", perfInfo.rating)
       )
       .foregroundStyle(by: .value("Performer", performer.name))
   case .area:
       AreaMark(
           x: .value("Cups of Coffee", perfInfo.cups),
           y: .value("Rating", perfInfo.rating)
       )
       .foregroundStyle(by: .value("Performer", performer.name))
       .position(by: .value("Performer", performer.name), axis:
   .horizontal, span: .inset(15))
   case .point:
       PointMark(
           x: .value("Cups of Coffee", perfInfo.cups),
           y: .value("Rating", perfInfo.rating)
       )
       .foregroundStyle(by: .value("Performer", performer.name))
   case .rect:
       RectangleMark(
   ```

```
        x:  .value("Cups of Coffee", perfInfo.cups),
        y:  .value("Rating", perfInfo.rating)
    )
    .foregroundStyle(by: .value("Performer", performer.name))
}
```

4. Lastly, in VStack containing the Text title and Chart, add Picker below it to switch between the marks:

```
Picker("Chart Mark", selection: $selectedChartMark.animation(.
easeInOut)) {
    Text("Bar").tag(ChartMark.bar)
    Text("Line").tag(ChartMark.line)
    Text("Area").tag(ChartMark.area)
    Text("Point").tag(ChartMark.point)
    Text("Rectangle").tag(ChartMark.rect)
}
.pickerStyle(.segmented)
```

Now, we can easily switch and see our chart with different mark styles, as we see in *Figure 12.5* and *Figure 12.6*.

Figure 12.5 – Our Bar and Line charts with various types of marks

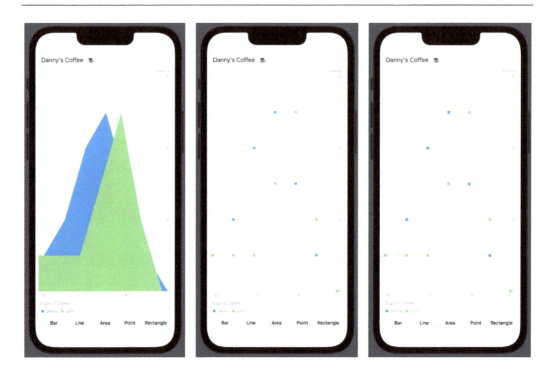

Figure 12.6 – Our Area, Point, and Rectangle charts with various types of marks

How it works...

Changing `Mark` of our chart is as easy as declaring a different `Mark`. It's as simple as switching `BarMark` for `LineMark`, or any of the other types! However, for this recipe, I thought it would be fun to be able to easily switch between the marks.

To do that, we created a simple `enum`, `ChartMark`, that would be the basis for our ability to switch. By declaring a `State` variable (`selectedChartMark`) as `ChartMark`, we can then inform `Chart` what state to be in. With these in place, we took two more steps.

First, we created a switch statement that takes in `selectedChartMark`. For every case that `selectedChartMark` can be, we then simply declared the corresponding `Mark`:

```
switch selectedChartMark {
case .bar:
    BarMark(
        ...
    )
```

```
case .line:
    LineMark(
        ...
    )
case .area:
    AreaMark(
        ...
    )
case .point:
    PointMark(
        ...
    )
case .rect:
    RectangleMark(
        ...
    )

}
```

Second, we added `Picker` with a list of `Text` and a corresponding tag for each type of `Mark`. This gives us a simple UI control on the user interface to change `Chart`'s state in our app:

```
Picker("Chart Mark", selection: $selectedChartMark.animation(.
easeInOut)) {
    Text("Bar").tag(ChartMark.bar)
    ...
}
.pickerStyle(.segmented)
```

There's more...

You may have noticed that some marks have a position modifier while others don't. That's OK! The `Bar` and `Area` marks are better served with that modifier. In fact, every mark can benefit from different types of modifiers.

For example, we can edit the `Bar` marks to have rounded edges, like in *Figure 12.7*, by adding the following code:

```
BarMark(
    // Existing code
)
.cornerRadius(10)
```

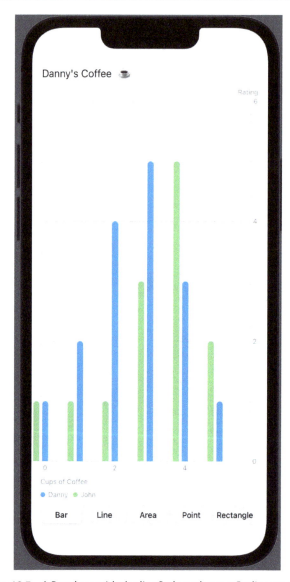

Figure 12.7 – A Bar chart with the lineStyle and cornerRadius modifiers

Or, we can change how the Line marks appear, like in *Figure 12.8*, by adding the following code:

```
LineMark(
    // Existing code
)
.lineStyle(StrokeStyle(lineWidth: 10))
```

Figure 12.8 – A Line chart with the lineStyle and interpolationMethod modifiers

A chart with an Area mark can take advantage of a gradient style, like in *Figure 12.9*, by adding the following code:

```
let curGradient = LinearGradient(
    gradient: Gradient(
        colors: [ Color(.red), Color(.yellow) ]
    ),
    startPoint: top,
```

```
        endPoint: .bottom
)

AreaMark(
    // Existing code
)
.foregroundStyle(curGradient)
```

Figure 12.9 – An Area chart with a gradient as a foregroundStyle

Or, we can change the `Point` marks into star symbols, like the ones in *Figure 12.10*, by adding the following code:

```
PointMark(
    // Existing code
)
.symbol {
    Image(systemName: "star")
}
```

Figure 12.10 – A Point chart with a symbol modifier

Lastly, we can mix marks together by declaring multiple marks in a chart. As a surprise, I'll introduce one more Mark called RuleMark, and we'll use it to highlight peak coffee performance. Add this below our switch statement:

```
if(perfInfo.rating == 5) {
    RuleMark(
        x: .value("Max Rating", perfInfo.cups)
    )
    .foregroundStyle(Color.pink)
    .annotation(position: .overlay, alignment: .leading) {
        Text("Peak")
            .foregroundColor(Color.pink)
    }
}
```

As we can see in *Figure 12.11*, it really helps point out our peaks!

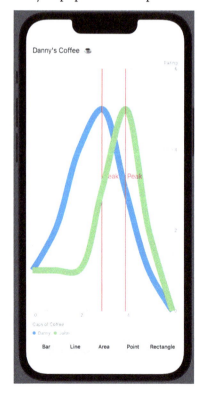

Figure 12.11 – A Line chart with an added RuleMark to highlight peak performance

There's plenty of customization and flexibility when it comes to Swift Charts. I encourage you to explore, bring in your own data, and have fun with charts!

Index

Symbols

3D models
 using, with ARKit 358-362
@AppStorage 295
@EnvironmentObject 295
@FetchRequest 296
@GestureState 296
.previewDevice 307
.previewDisplayName 307

A

access control
 used, for controlling access 89-102
advanced operators
 reference link 170
 using 166-170
anonymous functions 38
**Application Programming
 Interface (API) 195**
ARKit 351
 3D models, using with 358-362
 surface detection with 352-355
arrays 56
 data, ordering with 56-63

ARView 352
associated types
 reference link 166
associativity 179
Async/Await 227
 in Swift 247-250
 reference link 250
Augmented Reality (AR) 351
AVFoundation
 reference link 342

B

Bindings
 reference link 296
Bools
 using 7-14

C

CALayer
 reference link 350
chart
 building, with data 372-374
 marks and modifiers, exploring 379-388
class objects 27, 28
closed sourced 185

closures 38, 239
functionality, passing around 38-44
reference link 44
Cocoa Touch 252
Combine 287, 315
in SwiftUI 309-314
reference link 314
versus Delegate pattern 326-328
concurrency
Dispatch queues, using for 228-233
conditional unwrapping 106-108
CoreML
reference link 338
used, for detecting objects in
real time 342-349
CoreML models
used, for detecting objects in
images 334-337
custom operators
creating 172-179
reference link 180
custom types
subscripts for 76-81

D

data
chart, building with 372-374
fetching, with URLSession 190-193
ordering, with arrays 56-63
data flow
in SwiftUI 309-314
data, in sets
containing 63-66
intersection method 67
membership comparison 68-70
subtracting method 68

symmetricDifference method 67
union method 66
dates
comparing, with Foundation 186-189
decision making
with if/else statement 104-106
declarative syntax 288-290
default parameter values 17
defer statement
used, for doing it later 136-142
Delegate pattern
versus Combine 326-328
dictionaries 56
key-value pairs, storing with 70-75
Dispatch framework 228
DispatchGroups
leveraging 233-238
dispatch groups documentation
reference link 239
Dispatch queues
using, for concurrency 228-233
dispatch queues documentation
reference link 233
Domain-Specific Language (DSL) 290

E

enumerations 112
enums 32
associated values 37
computed variables and methods 35, 36
example 32
reference link 37
values, enumerating 32-35
error handling 124
catch block, using 126-131
do block, using 126-131

throw keyword, using 126-131
try keyword, using 124-126
eXtensible Markup Language (XML)
  working with 208-225

F

fatalError
  bailing out with 142-145
first in first out (FIFO) policy 232
floats
  using 7-14
Flutter 290
  reference link 290
force-unwrapped 74
for loops
  used, for looping 118-121
Foundation framework 121, 185
  dates, comparing with 186-189
frame 108
functionality
  encapsulating, in object classes 19-27
functionality, with extensions
  extending 86-89
function builders 290-295
functions 14
  code, reusing 14-17
  reference link 19

G

generics 59
  using, with protocols 157-166
generics functions
  using 154-157
  working 155

generics types
  reference link 154
  using 148-154
  working 152
Grand Central Dispatch (GCD) 227, 228
guard statement
  upfront, checking with 131-136

I

if/else statement
  conditional unwrapping 106-108
  decisions, making with 104-106
  enums, using with associated values 111
  optional unwrapping, chaining 108-110
image capture app
  building 330-334
images
  CoreML models, used for
    detecting objects 334-337
index 54
Instruments 341
Instruments Overview
  reference link 342
**integrated development
    environment (IDE) 2**
Interface Builder 254
interfaces
  defining, with protocols 44-47
intersection method 67
ints
  using 7-14
iOS 14 306
iOS app
  building, with UIKit and
    storyboards 252-274

J

JavaScript Object Notation (JSON) 311
  working with 194-208
Jetpack Compose 290
  reference link 290

K

key 70
Key-Value Observing (KVO) 83
key-value pairs
  storing, with dictionaries 70-75

L

Live Preview window 307
looping
  for loops, using for 118-121
  while loops, using for 121-123

M

Model View Controller (MVC) 273
multiple datasets
  displaying 374-379

N

name
  modifying, with type alias 81-83
nested types
  reference link 183

O

object classes
  functionality, encapsulating 19-27
  reference link 28

object-oriented programming (OOP) 19
objects
  detecting, in real time with CoreML 342-349
  detecting, in real time with Vision
    framework 342-349
Observable Objects 318
  using 319
  working 321
opaque return types 290-295
operation class
  implementing 239-247
Operation class documentation
  reference link 247
operators
  infix 176
  postfix 176
  prefix 176
OptionSet protocol
  reference link 172
option sets
  defining 170, 171
  working 171
OR operation 170

P

parameter overloading 18, 19
person occlusion 356
playgrounds 191
precondition
  bailing out with 142-145
Preview Provider 306-309
property changing notifications
  obtaining, with property observers 83-85
property observers
  used, for obtaining property
    changing notifications 83-85
property wrappers 290-295

protocol conformance 47, 48
protocol-oriented programming 326
protocols 44
    generics, using with 157-166
    reference link 49
    using, to define interfaces 44-47
publishers 321-326
pure function 16

R

reactive programming (Rx) 309, 315
Reactive Streams
    using 316, 317
    working 318
Reality Composer Pro
    using, for visionOS 362-369
RealityKit 355
Really Simple Syndication (RSS) 208
reference types 27
Resnet50 335
RxCocoa 309
RxSwift 309

S

SAX parser 211
set 63
SF Symbols 305
    reference link 309
States
    reference link 296
storyboards
    used, for building iOS app 252-274
strings
    using 7-14
structs
    reference link 32
    values, bundling 28-32

Structured Query Language
    (SQL) syntax 290
subscribers 321-326
subscripts
    for custom types 76-80
subtracting method 68
surface detection
    with ARKit 352-355
Swift
    access levels 89
    code, writing 2-7
Swift Charts 371
SwiftUI 287
    Combine 309-314
    data flow 309-314
    simple views, building 296-306
switch statements
    cases, handling with 112-118

T

test-driven development (TDD) 280
    stages 282
threads 228
tuples
    variables, bundling into 52-55
type alias
    name, modifying with 81-83
types
    namespacing, providing 180-183
    nesting 180-183

U

UIImagePickerController
    reference link 334
UIKit 251
    used, for building iOS app 252-274

UI testing
  with XCUITest 282-286
UIViewRepresentable 352
  reference link 309
Uniform Resource Locator (URL) 311
union method 66
unit and integration testing
  with XCTest 274-282
Universal Scene Description
     Zipped (USDZ) 362
upfront
  checking, with guard statement 131-136
URLSession
  data, fetching with 190-193
user interface (UI) 52

V

value 70
value-type semantics 30
variables
  bundling, into tuples 52-55
video capture app
  building 338-341
view controller object 254
views
  building, in SwiftUI 296-306
Vision framework
  used, for detecting objects in
     real time 342-349
visionOS
  Reality Composer Pro, using for 362-369

W

while loops
  used, for looping 121-123
Worldwide Developer Conference
     (WWDC) 1, 287

X

Xcode 1
Xcode IDE 251
XCTest 253
  integration testing with 274-282
  unit testing with 274-282
XCUITest 253
  UI testing with 282-286
XOR operation 170

Y

YOLOv3 342

www.packtpub.com

Subscribe to our online digital library for full access to over 7,000 books and videos, as well as industry leading tools to help you plan your personal development and advance your career. For more information, please visit our website.

Why subscribe?

- Spend less time learning and more time coding with practical eBooks and Videos from over 4,000 industry professionals

- Improve your learning with Skill Plans built especially for you

- Get a free eBook or video every month

- Fully searchable for easy access to vital information

- Copy and paste, print, and bookmark content

Did you know that Packt offers eBook versions of every book published, with PDF and ePub files available? You can upgrade to the eBook version at packtpub.com and as a print book customer, you are entitled to a discount on the eBook copy. Get in touch with us at customercare@packtpub.com for more details.

At www.packtpub.com, you can also read a collection of free technical articles, sign up for a range of free newsletters, and receive exclusive discounts and offers on Packt books and eBooks.

Other Books You May Enjoy

If you enjoyed this book, you may be interested in these other books by Packt:

The Ultimate iOS Interview Playbook

Avi Tsadok

ISBN: 978-1-80324-631-4

- Gain insights into how an interview process works
- Establish and capitalize on your iOS developer brand
- Easily solve general Swift language questions
- Solve questions on data structures and code management
- Prepare for questions involving primary frameworks such as UIKit, SwiftUI, and Combine Core Data
- Identify the "red flags" in an interview and learn strategies to steer clear of them

Test-Driven iOS Development with Swift

Dr. Dominik Hauser

ISBN: 978-1-80323-248-5

- Implement TDD in Swift application development
- Detect bugs before you run code using the TDD approach
- Use TDD to build models, view controllers, and views
- Test network code with asynchronous tests and stubs
- Write code that s a joy to read and maintain
- Design functional tests to suit your software requirements
- Discover scenarios where TDD should be applied and avoided

Packt is searching for authors like you

If you're interested in becoming an author for Packt, please visit `authors.packtpub.com` and apply today. We have worked with thousands of developers and tech professionals, just like you, to help them share their insight with the global tech community. You can make a general application, apply for a specific hot topic that we are recruiting an author for, or submit your own idea.

Share Your Thoughts

Now you've finished *Swift Cookbook*, we'd love to hear your thoughts! Scan the QR code below to go straight to the Amazon review page for this book and share your feedback or leave a review on the site that you purchased it from.

`https://packt.link/1803239581`

Your review is important to us and the tech community and will help us make sure we're delivering excellent quality content.

Download a free PDF copy of this book

Thanks for purchasing this book!

Do you like to read on the go but are unable to carry your print books everywhere?

Is your eBook purchase not compatible with the device of your choice?

Don't worry, now with every Packt book you get a DRM-free PDF version of that book at no cost.

Read anywhere, any place, on any device. Search, copy, and paste code from your favorite technical books directly into your application.

The perks don't stop there, you can get exclusive access to discounts, newsletters, and great free content in your inbox daily

Follow these simple steps to get the benefits:

1. Scan the QR code or visit the link below

https://packt.link/free-ebook/9781803239583

2. Submit your proof of purchase
3. That's it! We'll send your free PDF and other benefits to your email directly

www.ingramcontent.com/pod-product-compliance
Lightning Source LLC
Chambersburg PA
CBHW060649060326
40690CB00020B/4572